Google Classroom 2020

An Easy Guide on How to Teach Digitally in 2020 and To Manage Your Google Classroom Effectively

By

Ali Keler

The information herein is offered for informational purposes solely and is universal as so. The presentation of the information is without a contract or any type of guarantee assurance.

The trademarks that are used are without any consent, and the publication of the trademark is without permission or backing by the trademark owner. All trademarks and brands within this book are for clarifying purposes only and are owned by the owners themselves, not affiliated with this document.

Table of contents

INTRODUCTION...6

CHAPTER 1: INTRODUCTION TO ONLINE
CLASSROOMS 10

1.1 What is Online Learning?.. 13

1.2 History of Online Education 16

1.3 Different Types of Online Classrooms 20

1.4 Various myths about online learning 24

CHAPTER 2: ADVANTAGES AND DRAWBACKS OF
ONLINE LEARNING... 30

2.1 Advantages of Remote Learning....................................... 30

2.2 Limitations of Online Classrooms.................................... 38

CHAPTER 3: GETTING STARTED WITH GOOGLE
CLASSROOM... 46

3.1 Introduction to Google Classroom 46

3.2 Features of Google Classroom 55

3.3 Simple steps for setting up Google Classrooms 89

3.4 Adding and Grading Assignments.................................... 92

3.5 Privacy Evaluation for Google Classroom 98

3.6 Apps for creating content for Google Classroom 103

CHAPTER 4: GOOGLE CLASSROOM-AN INTERACTIVE PLATFORM 108

4.1 Engagement through student-teacher interaction 108

4.2 Engagement through student-student interaction............................112

4.3 Parental inclusion in Google Classroom 114

CHAPTER 5: GOOGLE CLASSROOMS-ADVANTAGES AND LIMITATIONS 115

5.1 Advantages of Google Classroom 115

5.2 Limitations of Google Classroom 120

CHAPTER 6: HOW TO GET THE MOST OUT OF GOOGLE CLASSROOM LEARNING? 123

CONCLUSION... 130

REFERENCES 131

Introduction

Education is essential for making one's life easier. It is undoubtedly the primary tool for improving one's life. A child's education starts at home. It is a lifelong cycle that does not end until one's death. Education undoubtedly defines the quality of a person's life. It strengthens one's awareness, abilities, and improves attitude and personality. Perhaps remarkable of all, education impacts people's chances of work. A highly educated person is likely to get a good job, as well. The drive for universal literacy is a phenomenon of the last 150–200 years. Traditionally, schools for young people have been accompanied by specialized training for priests, administrators, and experts. Monasteries associated with the Roman Catholic Church were the centers of education and literacy throughout the Early Middle Ages, preserving the selection of the church from Latin learning and retaining the art of writing. Until their formal establishment, many medieval universities were run for hundreds of years as Christian monastic schools, where monks taught classes, and later as cathedral schools, evidence of those later university's immediate precursors in many places dates back to the early 6th century. In the 6th and 7th centuries, the Gundishapur Academy, initially the Sassanid empire's intellectual hub and subsequently a Muslim learning center, offered medicine, philosophy, theology, and science training. The staff was not only versed in Zoroastrian and Persian cultures but also Greek and Indian studies. The education that we acquire today is certainly different from that of our primitive ancestors. The teaching methodologies, the concepts, and the means of providing education have evolved through time. The entire paradigm has shifted.

Initially, education passed down from the forefathers to their children-particularly the sons, so that they could carry forward the rituals and assist in their work. Their mode of learning was probably harsher, where they had to walk so many miles to get to their places of learning. Nevertheless, they had no option. Until today, physical classes have been prevalent. In this form of education, one has to leave their house and get to the educational institute to be taught. Although this form of learning could never go out of trend, because of some of its prominent benefits, it still is being replaced, to a large extent, by the online classroom learning. In online classes, one gets to learn at the comfort of their home. It is mainly due to the factors such as time and cost of traveling to faraway areas for learning, that the online classes have been introduced. Lack of attention in physical classes, potential bullying in schools and colleges, geographical immobility followed by a desire to learn various cultures being in a single place could be some other reasons. Live online lessons are similar in some respects to conventional face-to-face classes — an instructor can deliver information and communicate with a group of people in real-time — in other respects, some significant differences do exist. The first notable difference is the management of classrooms. The teacher is free to push the learners around in a physical environment, grouping them in various ways and setting out the class in a way that will ensure that the lessons run smoothly. Most online courses use conferencing software that allows for breakout groups and various learner configurations in the classroom.

Another distinction is the use of back channels. In educational contexts, such as seminars, back channels are especially common where the audience is forced to sit and listen for a more extended period. During the past, lecturers frequently banned cell phones from the lecture halls, but today, as a backchannel for the audience, progressive lecturers may also set up a twitter hashtag. Online classrooms usually have a text feature built into them, which the audience may take as a backchannel to ensure that they are engaged. Lack of visual feedback in online classes is a growing problem that teachers face. While talking directly to an audience in the same space, presentations can customize according to the audience's visual input. A room full of uninterested faces is a sure sign to a presenter or teacher that a more engaging activity needs to be added or their presentation- modified to engage the audience more. A cleverer presenter in the online classroom will make good use of the available resources to get similar feedback. Daily questions to the audience that they can answer using voting tools are a valuable way of ensuring whether the participants are participating in a live online session.

To sum up, although there are variations between physical and online learning, both delivery methods can be utilized in the hands of a well-trained instructor to accomplish the same goals. The chosen mode of teaching depends as much on the logistics of providing your training as on the subject being taught. However, live online lessons are becoming an increasingly apparent option for any professional development program in an increasingly globalized world. E-learning platforms have become even more crucial in the year 2020 when the world as a whole is stuck in the middle of a global pandemic, i.e., coronavirus, and people cannot leave their houses due to the lockdown situation. Under such circumstances, applications such as Google classrooms come to the rescue!

Chapter 1: Introduction to Online Classrooms

Online education is exhibiting an upward trend, specifically in the current year 2020. A recent study showed that 46 percent of new graduates take online credit as part of their degree, as more people shift to hybrid courses that combine distance learning with conventional classroom approaches than ever before. Most students are drawn to online learning versatility and combine their studies with work or personal commitments. Among some, a less costly alternative to conventional campus-based courses is distance learning. A three-year undergraduate degree will cost up to $50,000, while postgraduate courses can bring back around $5,000 for a student. And although online learning courses prices differ very significantly, comparable degrees cost around 50 percent less.

Many experts say the future of education is in online learning. If technology becomes more prevalent, more and more students can gain exposure to the kind of information that can significantly boost their job opportunities and maybe even enhance the planet. Online learning will allow students in developing countries to study topics such as coding, computer programming, and engineering and thus drive innovation around the globe.

Providing online courses and blended learning for many universities can play a crucial role in their long-term survival. Most students are covered by high tuition fees associated with full-time, campus-based classes.

Studying full-time challenges other essential responsibilities for others, such as a current career or raising a young family. After that, universities offering online learning services allow more students to enroll in their schools, including those residing in regional areas or, in some cases, in a different country. Any large organization must keep up with market demand, and rising technology and universities are no exception. Online learning, in other words, is not only here to stay; it is going to become an essential part of the future.

Besides, the online teaching approaches are already being introduced into the classroom by schools and higher education institutions, including innovations of the next decade, such as augmented reality, AI, and virtual reality. More than six million students have already had their VR-based classes, which involves virtual field trips to famous historical sites such as the Roman Coliseum, according to a tech-based study.

Meanwhile, companies around the world are developing with Google, a range of VR products that will allow virtual experiments to be performed by science students. One project involves a VR game in which a forensic science student will be able to examine and analyze the evidence they find in a virtual crime scene.

These innovations inevitably lead to wilder forms of speculation, but the emphasis on convergence rather than total replacement is essential to note in schools and big tech.

In other words, teachers in real life will always play a critical role in educating the next generation of students. They are essential.

Traditional classroom-based methods of teaching will still have a room, but even these are being combined with online learning.

And though predicting in which directions technology will take us in is almost impossible, online learning is expected to become a big part of educational institutions all over the world.

This approach is even more worthwhile in the current year when a coronavirus badly takes the world. Learning has been affected adversely, all around the world. About 185 countries have shut down their schools and universities, according to UNESCO. Education for hundreds of millions of children and young people has been affected. Education providers are trying to take up the task of teaching at a distance. Parents are working out their position, and teachers are doing their best in an ambivalent way.

Approximately 5.4 million South Korean primary, middle and high school students begin a single academic year in 2020 by distance learning, as the country aims to counter the COVID-19 pandemic.

One major problem is that certain places, colleges, and students are not equipped with the resources required to take part in these classes. Up to 85,000, students need devices such as tablets or laptops to take online courses, according to the Korean Ministry of Education. According to their ministry of education, it will purchase in excess, thus investing around 1.5 billion South Koreans won ($1.2 million) to upgrade some rural schools' wireless internet networks. The truth is that this is the moment of fame of online learning!

1.1 What is Online Learning?

Online learning is an education taking place over the Internet. In other words, it's sometimes referred to as "e-learning." Online learning, however, is only one form of "distance learning" - the paragliding term for any learning that takes place over a distance and not in a typical classroom.

Online education or distance learning is education for those students who may not always be physically present in a school. Traditionally, this usually included correspondence courses in which the student connected to the classroom by e-mail. Today, it is about online education. A distance learning program can either be ultimately distance learning, or a mixture of remote learning and conventional classroom instruction (called hybrid or blended). Massive open online courses (MOOCs), providing large-scale interactive participation and free access via the World Wide Web or other network technologies, are new modes of education in distance learning.

In a typical classroom, you also learn through listening, reading, writing, and doing other things that your teacher has designed. Online courses are different in the sense that the professor or other students are not in the same place. You'll probably never meet in person with your teacher or fellow students.

Students "attend" class in online courses by accessing the class web pages. According to the class schedule, they complete their assignments. Students use e-mail and online discussion boards to connect with the professor and their classmates. With busy schedules, this class structure is extremely versatile. Students will frequently log in to the course at any time of day (or evening). Being successful requires programming skills and determination. Students need typing skills and must be able to write to understand.

Difference between online and Distance Learning

Although all of these teaching approaches include the use of a computer or other gadgets by the students, there are some variations between them, i.e., online learning, as well as distance learning, need similar online learning resources, but the similarity ends there.

Generally, there are three main differences between online and Distance Learning:

- Location
- Interaction
- Intention

Location Differences

The primary distinction between online learning and distance learning is the location. Online learning or e-learning allows students to be with the teacher in the classroom while working through their interactive digital lessons and tests. However, students work remotely at home while the teacher assigns the assignments and digitally checks in while they use distance learning.

Interaction variations

Due to location variations, the relationship between teachers and their students often varies. Online learning can require daily in-person contact between teachers and students. This is because online learning, along with other teaching techniques, is used as a blended learning technique. Distance learning, however, involves no contact between teachers and students in-person. Nonetheless, you'll probably rely on digital communication channels such as messaging apps, video calls, discussion boards, and the Learning Management System (LMS) for your school.

Differences in Intention

The intent of the teaching technique is the final difference between online and distance learning. Online learning is intended to be used in conjunction with several other diverse methods of teaching in-person.

It's a supplementary way to change things up in your classroom and provide your students with a range of learning experiences. Distance learning is a technique for providing instructions solely electronically, not as a difference in the teaching style.

Each has a role in education at the end of day-online learning or distance learning. Depending on the needs of teachers and students, one will be better than the other. Online Learning works best for teachers in middle and high schools who want to provide their students with new forms of learning. Usually, distance learning fits better for older students who have constant exposure to technology at home and who work independently on their own.

1.2 History of Online Education

In 1728, one of the earliest attempts for remote learning was made public. It was in the Boston Gazette for "Caleb Philipps, Instructor of the New Method of Short Hand," pursuing students who wished to learn by weekly mailed lessons. In the 1840s, Sir Isaac Pitman offered the first distance education course in the modern sense, offering a method of shorthand through mailing texts transcribed into shorthand on postcards and collecting transcriptions from his students in exchange for rectification. The aspect of student feedback was a key advancement in Pitman's system. This scheme was made possible by the implementation of standardized postal rates throughout England in 1840.

This early beginning proved extremely popular, and the Phonographic Correspondence Society was established three years later to develop these courses more formally. The Society cleared the way for the later establishment of Sir Isaac Pitman Colleges around the country.

The Society to Promote Studies at Home, established in 1873, was the first correspondence school in the United States. Established in 1894, Wolsey Hall, Oxford was the first distance learning college in the UK.

Correspondence courses from universities

The University of London became the first institution to offer distance learning degrees, finding its External Program in 1858. The backdrop to this move was that the college (later known as University College London) was non-denominational, and there was an uproar against the "godless" university in light of the strong theological rivalries at the time. The question soon came down to which institutions had degree-granting powers and which ones did not. The adjusting solution that emerged in 1836 was that the sole authority to administer the exams leading to degrees should be granted to a new officially recognized agency called the "University of London." It would serve as an examination body for the University of London colleges, originally including King's College London and University College London, while granting degrees to the students for the University of London. With the State granting examination powers to a separate body, the groundwork was laid for the formation of a system within the new university that would conduct both exams and grant qualifications to students enrolled in some other institution or those following up self-study.

It was referred to by Charles Dickens as "People's University" because it offered access to higher education for students from less privileged backgrounds.

The External System was chartered by Queen Victoria in 1858, enabling the University of London to be the first one to offer remote learning degrees to students. Established in 1888 to train immigrant coal miners to become state mine inspectors or foremen, it enrolled 2,500 new students in 1894 and enrolled 72,000 new students in 1895. In 1906 gross registrations reached 900,000. The rise was due to the use of 1200 aggressive in-person sales clerks, sending out full textbooks instead of single lessons.

In the 19th century, education became a high priority, as American high schools and colleges greatly grew. Night schools were opened for people who were older or too busy with family obligations, such as the YMCA school in Boston, which became North-eastern University. Private correspondence schools outside the major cities provided a versatile, narrowly oriented solution. Large companies systematized their training programs for new workers. The Business Schools National Association grew from 37 in 1913, to 146 in 1920. In the early 1880s, private schools were opened across the country that provided advanced technical training to all those who studied, not just one company's employees.

Beginning in Milwaukee in 1907, public schools started to offer free training programs. In 1920, about one-third of the American population lived in towns of 100,000 or more; communication methods needed to be introduced to meet the remainder.

Australia was particularly involved with its large distances; in 1911, the University of Queensland established its Department of Correspondence Studies.

The University of South Africa, located in South Africa, previously an examination and testing body, began to teach distance education in 1946. The first meeting of the International Conference on Correspondence Education was held in 1938. The objective was to provide an individualized education for students at low cost, using a pedagogy of measuring, tracking, classification, and differentiation. Since then, the organization has been called the International Council for Open and Distance Education (ICDE) with headquarters in Oslo, Norway.

Open Universities

The United Kingdom Open University was created based on Michael Young's dream, by the then Labour Government headed by Prime Minister Harold Wilson. Planning started in 1965 under the Minister of State for Education, Jennie Lee, who set up an Open University (OU) model to expand access to the highest levels of higher education scholarship. A planning committee was set up, headed by Sir Peter Venables, consisting of university vice-chancellors, educationalists, and television broadcasters. Then, the British Broadcasting Corporation (BBC) Assistant Director of Engineering, James Redmond, had earned much of his training at night school. His inborn vigor for the project did much to solve the technological challenges of using television to transmit teaching programs. As well as research in other fields, it had been at the forefront of developing new technologies to enhance the distance learning service. In January 1969, Walter Perry was appointed first vice-chancellor of the OU, with Anastasios Christodoulou as its founding secretary.

The appointment of the Conservative government under Edward Heath's leadership in 1970 led to budget cuts under Chancellor Iain Macleod of the Exchequer (who had earlier called the concept of an open university a "blithering nonsense"). However, in 1971, the OU admitted the first 25,000 students, embracing a progressive free admission policy. Then, the total student population of traditional universities in the United Kingdom was around 130,000. Athabasca University, Canada's Open University, was found in 1970 and followed a similar trend, albeit independently developed. The Open University inspired the development of Spain's National University of Distance Education (1972) and Fern University, Germany. Some open universities have evolved into mega-universities, a word that denotes the institutions with more than 100,000 students.

1.3 Different Types of Online Classrooms

Training online comes in several ways. Modern classrooms are available and offer online activities. There are classrooms outside the school building, which are entirely online. And there are options for mixed classrooms which fall in between somewhere.

Internet technology has made many ways of distance learning possible through open educational tools and services, such as e-learning and MOOCs. Although Internet expansion blurts the borders, distance education technologies are divided into two delivery modes: synchronous Learning and asynchronous Learning.

All participants in synchronous learning are "there" at the same time. In this way, it parallels conventional teaching approaches in the classroom, given the remote position of the participants.

This needs setting a timetable. Web conferences, videoconferencing, educational television, instructional television are examples of synchronous technology such as direct-to-broadcast satellite (DBS), internet radio, live streaming, telephone, and online-based VoIP.

Web conferencing software helps promote meetings in distance learning courses and typically provides additional networking resources such as text chat, polling, hand radio. Such tools also facilitate asynchronous engagement by enabling students to listen to synchronous session recordings. Immersive worlds were often used to improve the participation of the participants in distance education courses.

Another type of synchronous learning that has entered the classroom over the past few years is the use of robot proxies, including those which enable sick students to attend classes. Some universities have started using robot proxies to create more engaging synchronous hybrid classes where students, both remote and in-person, can be present and interact using telerobotics. Remote students sit in front of a table or desk. They use telepresence robots instead of using a computer on the wall.

In asynchronous learning, learners view course materials flexibly according to their schedules. Students don't need to be together concurrently. Mail communication, which is the oldest method of distance education, is an asynchronous delivery system, and so are message board forums, e-mail, video and audio recordings, print materials, voicemail, and fax.

Many courses provided by both open universities and a growing number of campus-based institutions use periodic residential or day teaching sessions to complement distance-based sessions.

Recently, this form of mixed distance and campus-based education has been called "mixed learning" or less frequently "hybrid learning." Most open universities use a blend of technology and a variety of learning modalities (face-to-face, distance, and hybrid) all under the "distance learning" rubric.

Interactive radio and audio instruction (IRI) and (IAI), electronic virtual environments, video sports, webinars, and webcasts are often referred to as e-learning.

As for the forms of online courses, you can hear several different terms. Below are various concepts that may help standardize what is being referred to when looking at the many different online course choices.

Flipped course: Here, instructors record their lectures or use open-source tools to deliver content for the lecture. Students will be able to view the videos and complete a brief assessment of their comprehension of the material before attending class. A prominent benefit of this strategy is that at a time that works best for them, students complete the requisite training at their own pace and, the instructor acts as a facilitator during class, coaching, advising, answering questions, and dealing with misconceptions in real-time. The class will begin with a series of mini-lectures, further explaining material considered difficult or confusing by the students. The rest of the class is focused on active learning, and students and teachers connect. Many "flipped teachers" assume the most significant feature of the flipped paradigm is this. This classroom model is also referred to as an "inverted course."

Hybrid course: A large portion of the course learning process was transferred online in "hybrid" classes, allowing the amount of time spent in the classroom to be cut down. Standard face-to-face training is reduced but not eliminated. It is also called a "blended course."

Face-to-face: A traditional classroom setting where the students and the teacher meet synchronously in the same room; often referred to as "on-ground" or "on-campus" teaching.

Web-based: These classes are most often referred to as online classes, and the term "web-based" is more often used when referencing online classes. Online/Web-based classes typically use an LMS (Canvas, Blackboard, Moodle, etc.) to post instructors' syllabuses, materials, assignments, and communication. Most online courses are self-paced, and others have a unique test and task deadlines.

Web-Enhanced: In this specific course, there is a meeting in the classroom, which is much like a conventional class. There is a teacher at the usual every day, time, and place. Seat time is not replaced, so you will be able to navigate components of your course anytime using a Learning Management System (LMS) like Canvas. Online activities differ from teacher specifications and course specifications. Ideally, the professor will explain expectations on the first day of class, and students will have the flexibility to use an LMS, just as they do with any other technology.

Peer Instruction: Born in Harvard, Peer Instruction incorporates students during a lecture in their learning, and focuses their attention on the basic concepts. Conceptual questions, called ConcepTests, are interspersed with lectures, designed to highlight common difficulties in understanding the subject. Students are provided with one to two minutes to focus on the question and formulate their responses; they then spend two to three minutes debating their responses in groups of three to four, attempting to find consensus on the correct response. This method encourages students to think about the arguments being formulated and helps them to test their comprehension of the concepts long before they exit the classroom.

MOOC: A MOOC (Massive Open Online Course) is an online course designed for public participation on a wide scale and open enrolment. MOOCs are developed by university instructors and give students who are enrolled at the hosting institution a course credit. MOOCs are usually joint projects of institutions of higher education and consortia, including Coursera, Udacity, and edX. "Open" students can participate in the course activities and receive content delivered electronically while not receiving credit or direct input from the instructional team. Despite criticism of completion rates, MOOCs are thought to represent a new age of open learning, using network technology to promote interaction between scholar and student groups with no institutional or regional boundaries.

1.4 Various myths about online learning

With the increasing number of online college courses, you may be curious about your choices. Online Learning's simplicity and versatility are appealing, but is that format effective? You don't want the paper that is written on to be worth a degree; you want the skills and credentials that come with a great college program. Your questions can stem from some of the common myths about learning online. If there is some possibility that these misunderstandings are getting in the way of you and your dream future, it is time to bring them to light.

Myth#1 You have to teach yourself everything

Only because you're not physically present in front of your teachers, that doesn't mean they won't be there to provide you with assistance along the way.

Online class teachers have to engage their students and teach them the subject matter, just like in a brick and mortar class. Not attending classes in person can seem like a shortcoming (at first glance). However, professors tend to put extra effort into the instructions for the syllabus, arrangement of units, and assignments so that everything is directed to the maximum level. As in-person courses, when something is not obvious, it is up to the student to ask questions.

Myth#2 There is zero communication with class-fellows

You can utilize technology to speak to a friend face to face around the globe or work with a community of business associates, and you can even use it to communicate with teachers and classmates. Interaction is a fundamental part of most online programs where each student is usually required to post multiple times a week on course forums and occasionally meet with professors once at a time. Some professors and students think the posts at the website are much more interesting than just in-class conversations. Everyone has an opportunity to gather their views in writing and express themselves in their own time and in a more deliberate way than through contact in real-time. You would get a better feeling for who your peers are in a moderated online forum where everyone has to participate than in a lecture hall where only a few students speak up. Online conversations will give people a whole lot more insight. There are rules and etiquette on how to talk, and you'll quickly learn how people feel about a particular topic.

Myth#3 Professors are faceless

Professors often hold office hours in a typical brick-and-mortar university where students can stop by, say hello and

ask questions.

 They might also be available after class, or reachable by e-mail, depending on the instructor. Professor does a lot to strengthen the content and help students connect and understand. Professors can link to students via a Facebook program page as well as via their class platforms. The faculty can be very much available. Students could take full advantage of any chance of interacting through social media. You just never know what sorts of friendships both the message boards and social media can create!

Myth#4 You can't build a network

Talking of friendships, the people you meet are one of the undervalued benefits of education — and the connections they can bring into your life. Many people rely on college networks for landing jobs and learning about solid possibilities. That kind of relationship-building could seem hard in an online class at first glance, but the intentionality required in online platforms might make network-building easier. Although that is certainly not true of all online services, diligent counselors and managers (as well as instructors) should have a strong curriculum that can provide job guidance and connections.

Myth#5 Virtual classes are better than on-campus classes

When you are enrolling in an approved program, this is not the case! You may have been a little excited and optimistic reading this misconception that online education will be a better alternative for schooling. According to the professors, students should not only expect online courses to be just as demanding as their counterparts on campus but should also want them to be challenging.

The perceived distance from coursework online makes you feel as though you can put it off. If you're not a self-motivated individual, an online program may be especially challenging. Likewise, compared with conventional classes, online classes are not easier.

Online courses tend to be more challenging and take longer to complete. For student success in an online class, comprehensive reading criteria, and time management are needed for the assignment deadlines. Therefore, never underestimate the time commitment; keep pace and make sure you meet the demands of an online class!

Myth#6 To take online courses, you need to be a tech whiz

You can order a meal from your phone with only a few quick taps on the screen. You post the images and connect on social networks with friends and relatives. Hence, well-built technology doesn't have a learning curve down the steep. And online learning platforms are becoming more complex and equipped for better interactions with the students. Age or technological skill can never be a reason for you not to try learning opportunities online.

Myth#7 Online learning is an all-lecture

Video lectures are a part of many programs, but online learning goes far deeper than that. Online training does require a great deal of practical research. You'll be given tasks in specific online programs that mostly mimic what you're going to do in a real work setting—this offers you the opportunity to apply what you've learned and show your understanding.

Myth#8 Online degrees are not valued by the employers

When online learning was brand new, employers may have been wary—especially in industries where a degree is a non-negotiable qualification, such as education or healthcare. But online learning's popularity has been growing for many years, and what once seemed new is now very common.

The stigma against online degrees is fading as more and more employers, and those with the hiring authority have themselves received degrees online. There are several situations where an employer might also find an online education to be of higher value. Many of the updated technologies used in the offices are replicated in the virtual classes. Communicating and interacting online with colleagues is not unusual, so your online learning experience should prepare you for that interaction as well.

Myth#9 Online classes are all about procrastination

Procrastination does not go well with e-learning. Online class procrastination will cause more problems for the students than usual class procrastination. Students online need to be confident, empowered, and self-starters. Students must be able to set their schedules and observe them. The versatility of an online class needs to be handled by students.

Myth#10 All the work can be crammed into a single login session

Students usually have trouble being successful in a class when they only log in once every week or twice. Some students learn best through studying smaller quantities of content, then focus on the material before considering more. Besides, many teachers allow regular online discussion attendance. The dialogue not only helps students learn new ideas; in some cases, grade points are awarded for daily participation in the class. Grades will fail in many ways if students log in only once a week or maybe twice.

Chapter 2: Advantages and Drawbacks of Online Learning

Learning and education are considered a natural part of working and personal life. It should not be ignored both for achieving a job and for acquiring knowledge. The online world is continually evolving, and this provides a great learning opportunity. Discovering how to learn with all available communication channels and to select the ones that best match a person's filtering style of knowledge is very critical.

Online learning today is becoming more and more popular. Many mainstream universities started free, sharing of their courses online. It represents an easy and straightforward method for attaining knowledge in nearly every area, including law and accounting, human sciences, such as psychology and sociology or history. Online learning is an excellent alternative to traditional universities, especially for people who can't afford to take regular courses with the time and resources available to them. But what are the benefits and inconveniences of studying online?

2.1 Advantages of Remote Learning

1. Comfort: Online students should arrange their study time around their schedule. They will work when they have the most time, whether it is early in the morning or late in the evening. Students don't have to drive to campus, so they can save time and research anywhere they want. Sometimes, they don't have to make it into the library because the course materials are always available online, and they save more time. Students can also learn at their speed and research as quickly as they wish. Of these reasons, online education is the ideal choice of students wanting to balance their obligations to work and family. The most significant benefit of an online course is that you (theoretically) have your classroom and teacher available 24 hours a day, seven days a week. Isn't going online your only reason for missing the class? Otherwise, everything is there for you. You can get updates, access notes, review assignments, take quizzes for practice, answer questions, talk with fellow students, and study whenever you wish. Other than the various due dates, you set your timetable to complete the course requirements.

2. Flexibility: You may study whenever you wish. You can study with anyone you want to. Online courses allow you the freedom to spend time with work, family, and friends, significant others, or some other activity you want. You can study wearing whatever you like (or nothing if you prefer!). You will have to complete the job (and this flexibility can be your downfall). But for many people, online study options cannot be beaten. These people may be the ones who continually shift work schedules, or people making usual business trips. Parents with kids, students caring for others, or whose health keeps them from getting to the campus regularly and the students whose friends or boyfriend/girlfriend drop in unexpectedly, also fall in the said category.

3. Less expensive: Online programs are typically cheaper than traditional schools. Students also save on the related training costs. For example, students taking online classes do not have the costs of commuting or campus accommodation. They are also possibly not required to buy materials for the course, such as textbooks, which are freely accessible online. Some courses online are offered at no charge. For any computer that's linked to the internet, users will know. There is no need to rent a house or pay for high maintenance bills to provide a learning environment for the students. Furthermore, students and teachers are no longer going to drive to college, thereby saving them money too!

4. Diverse Courses & Programs: Online career colleges offer students several options. Students can find the online classes they need or degree programs – from nursing to neuroscience. They can also receive online every academic degree. As online learning trends upwards, more colleges offer students an option to take online degrees. Ashford University, for example, an accredited online university, provides diplomas in many disciplines, including business management, social and criminal law, early age learning, and studies of human behavior.

University Rivalry means that you are getting greater rewards. Online learning is rising, even as university enrolments continue to decline. Translation: Universities work very hard to draw students towards their courses online. Universities are finding that online degrees will bring students from far. Consequently, there is a lot of rivalry between universities in which you want to get enrolled!

Hence, there is a more fabulous option of universities for the students. You are not only restricted to the universities in the towns that surround you. You will be able to choose from a large number of colleges. For universities from Britain,

Australia, Canada, and New Zealand, you have the choice of any university that offers online degrees across your country. For American universities, you'll probably want to take advantage of government-based scholarships to go to a university in the state you're living in. Degrees online are evolving in leaps and boundaries. If you're signing up for an online degree, you're signing up for a kind, of course, that has a lot of money being poured into it by colleges fighting to earn the title of 'best online provider.' Universities will work very hard to keep you in their courses in exchange for your money.

5. Career Advancement: Online courses and degree programs, while seeking academic qualifications, encourage students to work. When working, taking online courses shows employers that you want to remain updated and can tackle new challenges. Penn State graduate, Kelsie Abduljawad, is a perfect example of someone who served while attaining an online degree. She earned a master's degree in leadership education, which she completed online via the Penn State World Campus. She worked at an all-girls Islamic school in Doha, Qatar, while she was receiving her degree.

You'll have experience using online learning resources such as Google Docs, Canvas, Blackboard, Webinars, Community Wikis, and Forums when you graduate with an online degree. To graduate with an online degree means you are a future-oriented, technologically competent employee of the 21st Century. You'll have a point of sale that the students on campus are less likely to make. This digital skillset would be at the forefront of your work application and could just get you on the shortlist to get hired on the interview day.

6. Greater Individual Attention: Since you have a direct pipeline via e-mail to the professor, you can have your questions answered directly. For fear of feeling stupid, many students aren't comfortable asking questions in class. The internet (hopefully) removes the anxiety (as long as the teacher feels comfortable). You sometimes worry about a problem after class, or when you're studying. Instead of trying to remember to inquire or forget about it, you should give the professor an e-mail. Your opportunity to know is heightened.

Following reasons depict why e-learning can be perfect for the introverts:

·   **E-learning does not involve the physical presence of a student**: The online world is a place where introverts can stand alone yet together. In such an environment, when learning will happen, introverts can rejoice. The thing about being around people has everything to do with enthusiasm. Extroverts get energized in a crowd, while introverts drain their energy. Most people do not know this is happening, but it does. In an online learning environment, both teachers and students are not in the same space as the student. It means that the introverted student will not waste their precious time paying attention to what's happening with other students in the class but instead will concentrate only on the lesson.

·   **E-learning is more self-paced:** Everybody learns differently, and every student advances in their way. But conventional teaching cannot consider this fact. All students will step forward at the same pace. Some who are going ahead need to wait for the others to catch up. Others who are struggling will have to deal with their learning gaps. Yet that isn't the case for e-learning anymore. Each student can grasp the concept at his or her own pace. Anyone who understands the learning materials can quickly finish the course in a fraction of the time. Those that need more time and additional clarification will get them to the same position without leaving anyone else stuck. For introverts, this lack of peer pressure is a great feeling.

·   **E-learning means more writing and less speaking:** Having an oral presentation is an essential mastering ability; no one can argue about it. Around the same time, though, this is not the only form of evaluation. Too many people talk a lot and still say nothing. Students also need to know that every word counts. Most of the classroom exercises include more talk and less writing, but the other way around is for e-learning. By default, introverted students like to write more than talk. When you're writing, every word has more time to process. And most of all, before you share it with others, you can edit your comments— strip out any unnecessary words or add more for clarification.

·   **E-learning offers more authority to the introverts:** Introverts are solitary creatures. Not only that, but they tend to figure things out on their own first — even the things that are harder to get — and they will only ask a friend or instructor for support when they meet a serious obstacle. They prefer the individual tasks of learning over group work. Because e-learning eliminates others' physical presence, introverted students can learn their way. Both the collaborative tools that a school LMS offers are fantastic to be available (and they'll use them!), but it's even better for these students to decide when to use them.

· **E-learning should go hand in hand with gamification:** The learning process never gets harmed by limited competition. But while most people compete with each other, the introverts compete with themselves. No matter how you look at it, it feels incredible to make noticeable improvements, and finally to win. As gamification elements can be incorporated so easily in online courses, competitive students will become more interested in what they study. Everybody's going to want to earn as many points as possible, add badges to their profile or get on the leader board. For introverted students, the most critical aspect of gamification may be the progress score.

7. It is modern: Today, most people tend to use the internet to access the content. Now, we're using the internet to read the news, watch our favorite television shows, and talk with friends, book appointments, shop, and more. Despite all the ease that the internet has brought to our everyday lives, why should education remain exclusively conventional rather than use the advantages that the internet offers?

8. Meeting Interesting People: Many of us, particularly in large classes, don't take the time to get to know our fellow students. Perhaps we're too distracted, or just plain shy. An online course provides the ability to get to know other students through newsletters, chat rooms, and mailing lists. Even if you're just chatting online, it offers you some kind of contact with other students and people that's just not realistic in the time-limited classroom on campus.

No technical innovation in man's history has connected the world with people like the internet. There is still a considerable variation between those who have Internet access and those who don't. The very fact that all of us can interact across the globe speaks about the value of this medium.

Several times in a course, the websites you visit are located in another country.

Can there be a better way to find out about Michelangelo's works than to go to Italy (of course virtually)? What better way can there be to learn about the Amazon rainforest or China's past or the customs of South Pacific islanders than visiting those places online? And if you engage in global learning days or other online activities, you can also catch up with someone in another country and make friends with them. After all, it is a small planet!

9. Self-Discipline: Procrastination is perhaps the greatest opponent of online courses. Some of the students, even teachers, put off the things which need to be done right up to the very last moment. If it comes to school, the last moment to know is the worst time possible. Often in the form of bad results at an exam or assignment, the lesson is learned the hard way. Yet, in the end, you are okay because you know the value of doing stuff on time or even in advance. That self-realization is what propels your online course success. No one is over your shoulder, waiting to tell you to go online and research. There is nobody there to get you to ask questions or post answers. You come with the ability to research in an online course. It is something student-centric or productive learning. The online student assumes responsibility for their study course and matures to become an adult for whom learning and achievement are highly valued. In short, that depends on you for your performance!

10. Promotes world skills and life-long learning: Understanding how to access knowledge online opens up a variety of chances for your personal and professional life. You can find online jobs, get online college applications, make online travel plans, and get online car dealer prices. You can also compare online shops, access great online works of art and literature, meet online people from around the world, follow online sports and movies, and so on! There are practically infinite possibilities. It gives you a definite advantage over someone who lacks those skills.

Usually, much of what we know in a course is forgotten at the end of the classes within a week or two. Finding that spark of curiosity and understanding how to find information online means you still have at your fingertips what your learning is. When you are interested in a specific subject, maybe because of something you see, read or hear about, or perhaps because you have a question from one of your kids or friends, you can get online and search it out. You will have developed the skills for finding information, digesting it, synthesizing it, and formulating an answer to any query that comes your way.

2.2 Limitations of Online Classrooms

Is there any limitation? Yes, online education does have certain drawbacks.

1. Excellent Time Management Skills Required: Believe it or not, online classes would take more time learning and completing assignments than an on-campus course does. How could this be? The education online is text-based. You will write notes, post comments, and otherwise communicate using your fingers (i.e., by typing) to connect with your teacher and other students. Typing is, as you probably infer, slower than talking. (Try to read each word as you type it, and compare the difference if you've spoken the same thing.) In the same way, reading the lecture materials can take longer than listening to the teacher delivering them, even if the lectures you've spoken have a distinct disadvantage. If you sit in a classroom, you will miss a large percentage of what the professor says, no matter how focused you are. For brief periods it is human nature to zone out. You may tend to go back through the notes as you're reading if you forget anything, and it takes more time. The point is you'll probably learn more in an online world, but you'll have to make a more significant effort to achieve that learning, and hence the time required for that.

An Internet-based course allows you to develop your time-management versatility. As in other words, if you don't utilize your time correctly, you will, of course, find yourself buried under a seemingly unconquerable coursework mountain. For completion of the studies, online courses require the self-control to allocate the time to study. This implies that learning online should be a priority, and one must not allow other things to intrude. Often, that means making tough choices.

2. Procrastination: Just as there is a dark side to that notorious property known as the Power, the Internet-based courses have a dark side too. This dark side starts with procrastination. Procrastination is an online course will rip you into parts. Nobody asks you to reach college on time. Nobody tells you that assignments are due, or that tests are coming. No one may preach to you, continue with you, and plead with you to keep on top of your coursework. (Sounds pretty sweet, huh?) Learning and assignments can be easily put off in the online world. Weeks have gone before you know it, you haven't done any homework, and it is exam time already. Timid. Anxious. Creepy. All too true.

It is a strategy for a sink or swim situation, and you cannot get it both ways. If you want to become this planet's responsible, self-sufficient, independently minded person, then now is the time to start. Life is not a rehearsal dress. Get into it!

3. Isolation: No-one will hear you scream in an online course. And this, for some online students, creates discomfort. It can be frightening to research alone with only the machine as your companion. There's no whispering in the back of the school, no wise peanut gallery comments, no imposing voice at the front of the classroom, begging everyone to listen. The online world is a very different atmosphere. Some people need to get used to it.

Your online teacher will hopefully be receptive to this issue and will help you resolve those feelings. In any case, if they begin to hinder your studies, you should be aware of them and seek support. A fast e-mail to a friend, your professor, or counselor will make you feel better connected if you're lacking the sense of community you're looking for.

Since the students taking online courses cannot communicate with professors and other students face-to-face, communication is by online chat groups or e-mails. Developing relationships with classmates in self-running classes is especially tricky. New schools, on the other hand, have a campus to socialize with other students or study with them. You may also pause to ask questions or get input from a professor's office.

4. Own Responsibility: It's just you who are responsible for knowing. No one can put that on you or get you to read. Teachers can only share some information and experience, give you a couple of tools, and hope you'll get it. You must have the spark and the drive to fulfill your dreams. So, the only downside of an Internet-based course, in a rational way, is that you do not own it. You may not be taking control of your research and your goals. You might well get a long way behind and never catch up. Online programs offer more flexibility for students, which can be challenging for students who don't know how to deal with it. Online courses often don't have teachers hurling you to remain on track, which means students are responsible for their learning and may not own it. It's easy to slip back and not feel inspired to catch on. There are no resources in the online classroom to help students learn so that the learning process can become more difficult. In the end, students need to be self-motivated to advance promptly through their courses and programs.

5. The problem for instructors: Online education is also a bit of a challenge for instructors. As tech progresses, teachers are continually trying to keep up. Traditional professors believe in lectures, so handouts and transition to the online course program can be difficult.

6. Technology Costs and Scheduling: The most critical elements of online courses are software programs and internet access. Students may need to learn new skills in programming and troubleshooting, which may take time. Students may also need to purchase new software to access their online classes or pay extra for upgrading to high-speed internet. Another drawback is that students need to change their schedules according to the due dates of assignments, which could be troublesome for international students or others who don't live in the same time zone as the teachers.

In the same way, if your laptop fails as a student on campus, you might need to get help from the campus library to complete your essays. As an online student, if your laptop crashes, you can't even finish your weekly tasks. You are in trouble. Problems with technology can emerge during your online degree. Be ready. Have you got a friend where you can use your laptop? Does your instructor record your live lectures so you can access them later? And, of course, are you confident learning how to use simple online tools such as Google Docs and Canvas, or management systems for learning Blackboard? You don't need to know how to use them right now, but you need the courage to learn in the short-term future how to use these devices.

7. Communication Breakdowns: Often, you'll give your instructor an e-mail and wait, and wait, and wait for a reply. Three days pass, and you eventually get a response that is, at best ambiguous.

So, you're e-mailing back, demanding clarity. Another three days pass, and you're getting another comment you just don't quite understand. Quickly, a week has gone by, and you're still trapped in the dark. That is the truth for a lot of teachers and students who aren't most updated with technology. Thankfully, this should only happen with your online degree once or twice. The downside of online degrees is that when it comes to technology, the teachers are generally very switched-on. But even a well-connected teacher could misunderstand the tone of your e-mail. Likewise, you could struggle with their language. The truth is, there is no contact form just as good as face-to-face contact.

8. The College Experience is missing: When you're an 18 – 24 teenager considering going to university, what are the key reasons you're going to study? Do you like to join clubs, make friends and go out to parties? Do you want to use the university to meet your potential spouse and to make connections for future jobs? Going to university is a great social experience. A degree that is entirely online does not give you any of the social coherence. Therefore, online degrees usually attract alternative students. They attract students who are of mature age, who work full time, or who have their own well-established family and social life and thus do not need or want the 'college experience.' You need to ask yourself whether you still wish to enjoy the university experience or whether your focus is to study versatility when continuing with your current life studying online.

9. Group Work can be challenging: Many university degrees nowadays impose group research as a prerequisite to passing the degree. The ability to work in teams is a preparation capability for the workplace that employer organizations believe is ingrained into a degree. When it comes to studying online, you are not other than needing to do group work.

Dreaded are the community of students who study online. The fact that they're interacting with somebody they have never met physically scares them. So they're reliant on their partners to log in continually. The trick with online community work is to try and locate a partner who will regularly and early in the week post on the forums. If you're working with one of those dedicated learners, you're going to do well.

10. Plagiarism and Cheating: Keeping in mind that students use a computer and are not always being monitored, they may plagiarize essays and other assignments. Along similar lines, cheating in online tests can be easier for students. Online cheating is easier to do (and more difficult to detect). Although it's not clear if online students are potentially cheating more than face-to-face students, the fact is that it's not easier to track who is taking a test and how they do it online than in a classroom. Nevertheless, the universities and professors on the opposite end may adopt some tactics to help counter the issue. Using technology for the detection of plagiarism, for instance. Having students run their essays through a plagiarism for-fee detection service will theoretically discourage piracy on cut-and-paste. At the very least, it could start a discussion about how to cite sources correctly.

Although e-learning comes with some drawbacks, the fact that the advantages far outweigh the disadvantages cannot be denied. You will develop self-discipline, and it is an attribute that can help learners in ways that go far beyond schooling.

In conventional learning settings, plagiarism and cheating may also occur, and there are ways to avoid that from happening in online tests that cannot be found in a typical classroom. Isolation can be overcome by integrating various learning approaches as in blended learning, which fosters further student interaction. Online education is an excellent by-product of the modern age. It gives thousands of individuals, who otherwise would never be able to pursue their studies for any reason, the ability to complete a study course. It is now totally up to the human race how it brings it into its most effective use-curse or a blessing.

Chapter 3: Getting Started with Google Classroom

Google has partnered with educators to create a classroom: a simplified, easy-to-use resource that helps teachers navigate the coursework. The educators will build classes with this Classroom platform, allocate tasks, rate and submit reviews, and see all in one place. Google Classroom is a straightforward, easy-to-use program, but you can learn a lot of best practices along the way. Join the Revolution in Google Classroom! It will completely change how you deliver assignments, communicate, and collaborate in your classroom and give your students skills that are ready for the future! Google Classroom saves you time and helps you to communicate with your students. Start today with resources, tips, and tricks from such educators like you. Set up your classroom for success, and be prepared to be amazed at the ease and simplicity that Google Classroom brings to your workflow.

3.1 Introduction to Google Classroom

What is Google Classroom?

Google Classroom is free of charge web service made by Google for schools. It clarifies paperless structure, dissemination, and marking of homework. Google Classroom's underlying goal is to streamline the file-sharing process between teachers and students. Google Classroom incorporates Google Drive for the production and delivery of tasks, Google Docs, Sheets and Slides for writing, Gmail for collaboration, and Google Calendar for scheduling.

Students can be invited via a special code to enter a college, or imported automatically from a school domain. Every class creates a separate folder in the Drive of the respective individual, where the student can send work for a teacher to assess. IOS applications, which are available for IOS and Android devices, allow users to take images and add to assignments, share files from other phones, and offline access. Teachers can track each student's progress, and teachers can return work along with the feedback after grading.

But what distinguishes Google Classroom from the standard Google Drive experience is the interface between teacher and student, developed by Google for the way teachers and students think and work.

Evolution of Google Classroom

Google Classroom was revealed on 6 May 2014, with a preview available to individual members of Google's G Suite for Education program. It was launched publicly on 12 August 2014. By October 2015, Google reported that certain 10 million students and teachers were using it. Google said about 50 million students and teachers worldwide used Google software, from Gmail to Chrome.

In 2015, Google introduced a Classroom API and a website sharing button allowing school administrators and developers to continue their interaction with Google Classroom. In 2015, Google also incorporated Google Calendar into the Classroom for planned assignment dates, field trips, and class speakers.

In 2017, Google enabled Classroom to allow any personal Google users to enter classes without the need for having G Suite or Education account. And it became possible for any individual Google student to build and teach a course in April of the same year.

In 2018, Google announced a refresh classroom, introducing a classroom section, enhancing the grading interface, allowing teachers to reuse classroom work from other classes, and adding features to organize content by topic. In 2019, Google released 78 new illustrations of the classroom.

For the previous two years, Google has been taking its popular apps and equipping them for classroom use. Although many schools and districts tend to use traditional learning management systems, such as Blackboard, Canvas, Moodle, and Schoology, the eyes of teachers are gradually focusing on Google's Classroom Platform. Most schools are also using the collaboration software suite of Google — Docs, Sheets, and Slides. What Classroom seeks to offer is a way of bringing together these applications and applying new functionality to what teachers and students need. In short, the Classroom is aiming to be a lightweight framework for learning management.

According to the product manager at Google, they spent about a year and a half studying and talking to educators about the app. Apps alert from the guardians and the introduction of multiple teachers to a class were created merely from user feedback.

Does Google Classroom become an LMS?

Technically, it doesn't. Google Classroom is not a stand-alone program for learning management (LMS), course management (CMS), or student information (SIS) program. That said, Google adds new functions to Google Classroom periodically. For example, in June 2019, Google announced that schools would soon be in a position to synchronize the new grading features of the tool with an existing student information system. As Google continues to add features, it is likely to start looking, becoming more like an LMS, to work. Perhaps it's better, for now, to think of the device as a one-stop-shop for class organizing.

Is Google Classroom free of cost?

The Google for Education platform is free for schools. Still, there is a paid G Platform Enterprise tier for education, which includes additional features such as advanced video conferencing apps, advanced security, and premium support. Google no longer publishes information about pricing, so you'll want to contact them directly for a quote. Google also offers several free items for authoring tools, web themes, and professional growth, such as Chromebooks, and partners with other companies.

Implementation and integration

Google provides educators and IT Administrators with a range of training choices. These are:

- The Teacher Hub, which provides primary or advanced self-paced Google Classroom preparation and instructor professional development resources

- Train the Trainer course for people into teaching others

- Google software Certified Educator and Certified Instructor programs

- G Suite Certified Administrator program for IT administrators

Who is eligible for Google Classrooms?

·The classroom is open to:

·Schools that use G Suite for Education

·Organizations that use G Suite for non-profits

· Individuals over the age of 13 with personal Google accounts. Age can vary according to region.

·Both Domains in the G Suite

Google Classrooms Support Service

Users can access Google Classroom help in the following ways:

· The Help Centre offers information on different topics related to Google Classroom. There is also a troubleshooting section with solutions to common issues.

· There is a software community where users can seek advice from other Google Classroom users and Google Classroom staff.

· Google Classroom also offers regular updates with new features and other software enhancements.

·IT guides for schools' IT administrators also exist.

Can the Classroom be used if G Suite for Education domain includes Gmail disabled?

Yes. Gmail doesn't have to be enabled to utilize the Classroom. If your administrator has not activated Gmail, however, teachers and students do not receive e-mail notifications.

Note: If you set up your mail server and receive information from Drive, you can receive notifications from the Classroom too.

Could the Classroom be used if G Suite for Education domain has been disabled?

No. Classroom collaborates with Drive, Docs, and other tools offered by G Suite for Education to help teachers build and collect assignments, and students submit work online. If you disable Drive: Docs and other services are also disabled. You are not able to add these resources to the research that you allocate to students. Students would also not be able to add these to their jobs. The classroom can still be used, but the collection of features is minimal.

Difference between Google Classroom and Google Assignment

Google Assignments is for organizations using a learning management system (LMS) that want better grading workflows and assignments. It can be utilized as a stand-alone tool and a complement to the LMS, or it can be implemented into the LMS as an interoperability learning tool (LTI) by the school admin. If you are using the Classroom, you already enjoy the best of tasks, including reports on originality.

Accessing Google Classroom from school account and personal account

Most of the time, the classroom is the same for all users. However, since users of school accounts have access to G Suite for Education, they get further attributes, such as email summaries of student work for guardians and full user account management. G Suite for charity users has the same features as users of G Suite for Education.

Google Classroom for visually impaired people

Google Classroom is a resource designed to help teachers and students interact in the paperless classroom and remain organized there. Visual disability and blindness students can use a screen-reader to access and handle classes and assignments.

Following is the availability for various screen readers:

Web: With any modern browser such as Chrome, Mozilla Firefox, Microsoft, Internet Explorer, or Apple Safari, you may navigate the Classroom using a screen-reader. See the guide on how to set it up in your browser. You can use ChromeVox, for example, with your Chromebook. On Macs, the built-in screen reader, VoiceOver, is used.

Mobile

Android: The smartphone app for the Classroom works with TalkBack, a pre-installed screen reader that uses spoken input for interaction.

IOS: The Virtual Classroom app operates on IOS with VoiceOver. For specifics, you might need to see your device's accessibility settings.

Google Classroom API outline

The Classroom API can be used by schools and technology companies to create applications that communicate with Classroom and G Suite for Education and to make Classroom function better to suit their needs. The Classroom API is an API created by Google. That means non-Google companies will benefit from the resources and infrastructure that Google provides.

To use the Classroom API, developers must adhere to the Terms of Service of the Classroom API. Many programs cannot use Classroom data for marketing purposes. Third-party developers and administrators may use the Classroom API. Teachers and students must approve third-party apps. Utilizing the Classroom API, you can do many of the things that teachers and students can do programmatically through the Classroom UI. For example, you can synchronize with the student information systems, display all the classes taught in an area, and control the coursework.

Non-Google services can use the Classroom API to incorporate Classroom features. For example, an app may allow a teacher to copy and reuse a Google Classroom class, rather than re-create the level and re-add every student. Applications may also display, build, and change Classroom work programmatically, add materials to work, turn students' work in, and return grades to the Classroom.

The software must request authorization from the Classroom user before software or service can access Classroom data. The app asks for the individual permissions it requires (such as a username, email address, or photo profile), and the user may approve or reject the request made by the service. The

Classroom API uses a popular Internet standard named OAuth to authorize access.

As an administrator of the G Suite for Education, you monitor how the data is exchanged within a domain. You can decide which teachers and students in your area can allow services to access their Classroom data in the Google Admin Console. By organizational unit, you can customize the access. You can also monitor the services that have been given access to a user's account in your jurisdiction in the Admin console, and you may revoke permissions if required.

The different tasks that the Classroom API can accomplish depend on what position a user has in a class. A user can be a student, instructor, or administrator just as in the Classroom UI. Teachers and students should accept applications from third parties and report misconduct.

When the consumer is a (n):

Student: The API can view the course information and teachers for that course.

Teacher: The API can build, display, or remove their classes, show, attach, or remove students and additional teachers from their classes, as well as view and return research, build assignments and topics, and set grades in their classes.

Administrator: The Classroom API can organize, view, or delete any class in their G Suite for Education domain. It can attach or delete students and teachers in their area in all the classes. It also looks at the work and topics in all of the classes in their domain.

There are many explanations for why Google Classrooms are being used for more and more classrooms. The technology is being implemented all around the world, including the US schools and districts through the 1:1 laptop initiative. The initiative features a learning laptop for every pupil. Chromebooks are often chosen because of their affordability and intuitive interface. They are easily integrated with the complete suite of Google apps that includes the Classroom.

The basic requirements to get started with the Google Classroom application are to get a device or gadget that would render the installation of application possible. Such devices may include a smartphone, tablet, laptop, or even a desktop. This would then be followed by access to a useful quality internet with fast speed and more MBs. Users would need to ensure that there are no power-cuts or any other issues which might interrupt the operation of their Google Classroom. It is, however, to be noted that Google Classroom application would only work in those geographical locations which have the internet services available. Furthermore, Google Classrooms provide course accessibility in various languages used in different corners around the world. These languages can be changed by adjusting the settings inside the application. A more detailed guide regarding Google Classrooms would be provided in the forthcoming topics of this book.

3.2 Features of Google Classroom

As the classroom is increasingly paperless, teachers need to start seeking strategies for handing out tasks, handling their classrooms, engaging with students, etc. An increasingly growing number of teachers find their way into Google Classroom. An innovative immersive classroom with less emphasis on software and more emphasis on teaching. You needn't be a trained tech to manage this classroom.

Wouldn't it be awesome if you could arrange your students' tasks, resources, and grades in a single location? Thankfully, Google listened to the teacher's needs attentively for EDU and planned Google Classroom to do that. It can be defined as the anti-LMS since it is both easy and efficient. Students have a common source of information for assignments, parents can see missed assignments and progress for the students, and educators can handle digital assignments and communication more easily.

Google Classroom allows teaching to be more effective and meaningful by providing educators with a forum for student assignments, fostering collaboration among students, and promoting communication. Educators can create classes, distribute tasks, submit input from individuals, and see all in one place. The classroom also integrates easily with other Google resources such as Calendar, Google Docs, Photos, Drive, and others.

And the most relevant issue is undoubtedly this. How should you use the Classroom on Google? What's for you in it? It's completely safe, first of all. You're not going to need to upgrade to a pro edition, which will save you some money. Sure, $0.00. Nothing. You can get started after you have configured your classroom. Here's a rundown of what the prominent features of this wonderful application are.

Add announcements and lesson material: Offer advertising about your lesson to your students. The announcements include the lesson supplies. Such announcements will appear in the Google Classroom stream of your students. This way, the students will easily find anything. You can attach materials from a Google drive, connect to that lesson in Google Classroom, add files and pictures from your phone, add a YouTube video, or add any other connection that your students want to see. That is so simple!

Add assignments: You can add an assignment to your course just as you add an announcement. It operates the same way except you get the option of adding a due date and rating it here. When they have to make an assignment, it will alert the students, and it will also appear in their calendar.

Marking or grading an assignment: You can then review and rate the assignments the students have sent in. There's room for feedback via a comment from a teacher. Instead, return the task to your students. The "Points" tab houses a grade book of the assignments and grades of the students.

Manage students: The students must, of course, be able to share their thoughts. Or don't they? That is entirely up to you! You can manage permissions, allow students to post and comment, comment only, or give the teacher the ability to post and comment only. Even the students can be e-mailed individually.

Post Questions: You can post questions to your classroom and allow students to have discussions by answering each other's responses (or not, depending on the setting you choose). For instance, you might post a video and have students solve a query regarding it, or pin an article and ask them to write a response paragraph.

Reuse the Assignments: If you reuse your curricula year after year – or at least reuse papers, you may want an update. You can now recreate assignments, announcements, or questions from any of your classes — or from any class, you co-teach, be it from last year or last week. If you pick what you would like to copy, you will also be able to make changes before publishing or assigning them.

Better Compatibility of Calendars: Users prefer improvements that boost workflow. The classroom will automatically create a calendar in Google Calendar for each of your classes over the next month. All tasks that have a due date will be added to your class calendar automatically and kept up to date. You can display your calendar from within the Classroom or on Google Calendar, where you can add class activities such as field trips or guest speakers manually.

Bump a post: Sticking posts have long been a feature on forums, tweets, or Facebook updates. Now you can also do it on Google Classroom by pushing every post upwards.

Optional Project-based learning due dates: Self-directed learning? If you're using long-term projects or other tasks with no due date, you can now build tasks in Google Classroom without due dates.

Add a Google Form to a post: If you're a fan of Google Forms (here's a post about creating a self-graded exam using Google Forms), this is a movie you'll appreciate. A lot of teachers have used Google Forms as a simple way to allocate the class a study, quiz, or survey. Teachers and students will be able to add Google Forms from Drive to posts and assignments and get a connection in the Classroom to see the answers quickly.

YouTube Features: Love YouTube, but is it about inappropriate content? Google listens. As it also includes content that an organization or school does not find suitable, Google Classrooms introduced advanced YouTube settings as an Additional Feature last month for all Google Apps domains. These settings enable Apps admins to limit the viewable YouTube videos for signed-in users, as well as users who are signed-out on admin-managed networks.

Google has announced several changes to its Chromebook Device Platform. The app shop built for educators offers not only learning and development applications but a multitude of "ideas" to help motivate instructors to make the most of the technology in their classrooms. The latest customization to the Chromebook App Hub will make it even easier for teachers and administrators to use the latest Chrome OS hardware that students are using.

Drag and Drop: On the Classwork Page, Google rolled out the new Classwork page last fall, where teachers could stay organized and chart their classes. Nevertheless, teachers organize their classes in various ways and require more versatility in their resources in the classroom. So now, you can drag and drop whole topics and individual items in the Classwork, easily rearranging them on the list. On the Classwork tab, you can drag a whole subject to a specific location, or drag individual items into — and in between — themes. This feature was launched on mobile last year, and now it's time for it to hit the web.

Refreshed Ux: Starting in January 2019, users were also able to see that the Classroom had a better and new look and feel, first on the web, and soon on the mobile apps in the classroom. The Company launched Google's latest content theme back in 2014 to provide greater consistency across Google's products and platforms. You'll see a more fluid style flow among the changes — plus a new approach to form, color, iconography, and typography on both the web and the mobile app. This also makes the class code easier to access and project so that students can find and enter easily.

And finally, 78 new themes were launched with design illustrations, ranging from history to math to hairdressing to photography. Now, more than ever, you can customize your Classroom.

Improved Training & Help: The need for more help comes with new resources and improvements. In the Teacher Centre, you could find the revised videos with the latest template and functionality that were rolled out in 2018 on the First Day of Classroom Training. New and enhanced Help Centre was created while Google support was at it, in tandem with the Group and product platform.

So what is new?

With tablets, convertibles, and touch-enabled devices increasingly trending in the classroom, new specifications in the Chromebook App Hub will let teachers find apps designed for these form-factors quickly. Here are some examples from The Keyword:

- Search for your favorite apps and ideas and share them with other educators.

- new filter options that enable teachers to search for the best app to improve their Chromebook tablet lessons by topic

- software feature and Google integration, and the ability to filter apps through privacy laws such as GDPR and COPPA

The last filter has allowed more than 20 applications to be updated already to take advantage of Chrome OS tablet mode, which will support users such as the newest Chromebook tablet Lenovo 10e.

Below is a detailed outline of all the updated features Google has introduced so far, over the last few years.

April 2020

New feature: Video class meetings — Teachers can start for distance learning, and students can enter video meetings in Classroom with Google Meet.

Only teachers in the Classroom can build video-meetings. Both video meetings produced in the Classroom are called nicknamed meetings, so if the teacher is the last person to quit, students cannot start a session before the teacher, or enter the meeting. These permissions will vary according to how the school administrator sets up Meet.

To use Meet in the Classroom: School accounts must be used by teachers and students and be in the same domain. Admins have Google Meet to turn on. Admins can find further support for distance learning in Set up Meet.

Additionally, the new G Suite Enterprise for Education capabilities will be available free of charge to all Classroom educators using G Suite for Education by September 30th,

2020. Such apps include live streaming, recording, and video meetings for a class of 250 students. Live sessions are available on the Classroom's Web and Mobile versions.

January 2020

New features: Notifications of originality — Teachers can now turn on three assignments per class for originality results. The reports highlight source material for students in their research and flag missing references so they can improve their writing. Teachers will display reports after the students send work to check academic integrity and provide input from the grading tool. With G Suite Enterprise for Education, administrators can upgrade to unlimited originality reports.

Rubrics — Teachers can now build rubrics and reuse them. After the completion of work, students could review the rubric of an assignment to help them stay on track. As teachers rate rubrics, level choices will automatically determine a cumulative grade, which can also be manually modified. Students will quickly check their rubric feedback upon returning to work.

All about rubrics: Customizable — up to 50 requirements and ten-value standards Shareable — Import and export options when you build interchangeable assignments — **Reuse a rubric in another assignment Note:** Rubrics roll-out course by course, and should be available in a few days.

New mobile features: See overall grades — if teachers share whole categories, students can see the overall grades on a mobile device (Android and iOS).

See rubrics — on a mobile device (Android and iOS), teachers and students can see the rubric of a task.

New beta programs: School matches — Expands originality reports inside the school to check for events. For information, go to Sign up for Beta programs in the Classroom.

Internationalization — Reports of originality appear in multiple languages. For info, go to Sign up for Beta programs in the Classroom.

August 2019

The new smartphone features: Show beta rubrics — Teachers and students can now see rubrics on their Android device or IOS.

Enhanced IOS app — Students would press "Your job" to show their files at an assignment.

New updates: Classroom with the Classwork page- Google launched a new Classroom edition with additional functionality in August 2018, including a Classwork page to help teachers coordinate classwork. Teachers may also return to the previous Classroom edition.

The last version (without the Classwork page) was deprecated on September 4, 2019, and discontinued. As a result, teachers were no longer able to have the option of deleting the Classwork page or reverting to the previous edition as of September 4, 2019. Any classes that were using the previous version were migrated automatically to the new version beginning September 4, 2019.

For teachers, what does that mean?

Any classes they had in the previous version were migrated automatically to the current version beginning September 4, 2019. Class resources were not transferred to the transformed courses in the Class Settings tab. Such materials (excluding YouTube content) could still be accessed in the teachers' Drive tab. By adding the same stuff to the Classwork list, they could create a similar experience

If you, the teachers, were using the previous Classroom version, you were enabled to move it to the new version by adding the Classwork page before September 4.

What would admins do?

Although instructors were informed of this change through in-product notifications beginning in August, Google recommended notifying instructors of this update in the domain.

New features for August 2019: Beta originality reports — you can now sign up for beta originality reports. Teachers should turn on stories of originality while they are making an assignment. For more stuff, go to Sign up for Beta programs in the Classroom.

June 2019

New features: Archive a class — Teachers are now able to archive mobile device classes on IOS.

Redesigned student assignment page — Students are more likely to apply research and communicate with their teachers than ever before. Go to Request an Assignment for details.

Gradebook: The Grades page — Teachers will record grades from the Grades page and return them.

Grading systems — Teachers can select one grading system for each class.

Grade Categories — Teachers may assign classwork posts to grade categories.

Overall grade — if an instructor wants, students will see a class for their overall score.

Docs grading tool — Teachers may receive input from the Docs grading tool and grant grades.

New Beta programs: Link grades to your SIS — Teachers can transfer classes to their student information system (SIS) directly from the Classroom. To voice interest, administrators can go to the sign-up page for beta interest.

Rubrics — Teachers can build and save custom rubrics for assignments to grade and share feedback with students. Teachers or administrators can go to the Beta sign-up page for rubrics to show interest.

May 2019

New apps for Android: New tablet grading view — Teachers can apply grades on tablets to a list of submissions on the left, and give the individual student feedback on the right.

The Preferences tab has a fresh look.

New IOS feature: Student Selector — Teachers can randomly select students to use the Student Selector to call.

Randomly pick a student: This information is mainly for teachers. Teachers can use the Student Selector of Classroom to call on their pupils randomly. The student selector selects students randomly from the class roster. Teachers could call a student, skip a student to call later on, or mark an absent student. This functionality is only available on mobile devices running Android and IOS.

Note: The Student Selection is neither accessible nor visible to the students.

The following steps might help:

To select a student

· On your mobile app, pick a student and tap Classroom icon, then Classroom, then your class, and then People.

· Tap Student selector icon, then Student in the top-right corner

· **Choose one:** Name the student shown and then tap "Next" for another student.

· You can switch to the next student on iOS devices too.

· Tap Call Later, to skip the student display.

· Tap Absent to identify the shown student not present for the selection session.

· (Optional) Tap Start Again at the end of the meeting if you wish to reset.

To see the class roster

· To see a list of those who were marked as Selected, Not Chosen, or Missing, go to the roster.

· Tap Classroom icon, then Classroom on your mobile app, later on, your class and then on People

· tap Student selector Student icon, then selector in the top-right corner

· Tap Not Selected, Picked, or Absent under their numbers.

· You'll see which students haven't been chosen, picked, or absent.

· To get back to the Student list, tap Open Back or Close option in the top-left corner.

Reset the student selector

- Click on the Classroom icon, then Classroom on your mobile device and later on your class and then on Students.

- Tap Student selector option and then selector in the top-right corner

- Tap Reset, in the top-right corner

- Tap Reset to check again.

- All students are indicated as Not Selected.

April 2019

New features April 2019: New work now posts to the top of the Classwork list.

Teachers will filter the Classwork page by theme now.

New features for Android: Co-teachers can now leave a class on the People's list.

An instructor can be inviting you to teach as a co-teacher in their class. Once they enter a class, co-teachers will perform all of the instructor's tasks. Primary teachers and co-teachers, however, in Classroom have separate permissions.

Such essential distinctions are:

· The primary teacher can only delete a class.

· The principal teacher is unable to unwind or be excluded from the course.

· Teachers are not permitted to be silenced in a class.

· The primary teacher manages the Google Drive class folder. The co-teacher is granted access to the class Drive folder after a co-teacher enters the class.

· Your G Suite administrator could just encourage your school teachers to begin classes. If you have trouble adding yourself to a class, please contact your admin to change the class membership settings for your domain.

· If you quit a class that you co-teach, you cannot open it again unless you are re-invited or enrolled in the class as a student.

Accepting an invitation

· Head to classroom.google.com to approve a request, and press Sign in.

· Log in to your Google Account. For instance, you@yourschule.edu or you@gmail.com.

· Select Allow on the class card if you want to teach the lesson. If not, please press Decline.

· **Note:** If you're a college member, it does not delete you from college by clicking Decline.

· (Optional) The invitation can be accepted by clicking on the link in the invitation document.

Leaving a class

You have two ways to quit a class as a co-teacher.

From the list Classes:

· Go to classroom.google.com and then press Sign In.

· Sign in to your Google Profile. You@yourschool.edu, for instance, or you@gmail.com.

· Click the "More" option and then Leave Class on the page you want to quit.

· Tap "The Leave Class" option to confirm.

From the People page:

· Select More and then Leave Class across from your name.

· Click on Leave Class to confirm.

Share files with the primary teacher before leaving a class. The primary teacher will access all the data you have generated in the class in the class Drive folder. You can link them to the class Drive folder if you want to share other files, such as notes or an attendance sheet. The data or directories you add to the class Drive folder can only be accessed by teachers and co-teachers. When you share them, students cannot access the files or directories.

· Go to the.google.com classroom, and press Sign In.

· Sign in to your Google Profile. For instance, you@yourschool.edu or you@gmail.com.

· Select the Open folder on the class card.

· Click New in the top-left corner and then select an option: Folder — to add a new folder for your files to the class Drive.

- Import file — to access a folder that was developed beforehand.

- Upload of the folder — to delete a folder that was generated before.

March 2019

New features: Improvements to private comment

notifications — Private comment notifications can now be switched on or off. You will also obtain them separately from other updates about the job. By default, you get email alerts for a range of things, such as when someone comments on your post or returns work to your instructor. These notification settings can be changed at any time.

Students and teachers are permitted to:

· Turn on or off all alerts.

· Select what updates you'll get.

· Turn on or off alerts for class.

· Turn on or off email alerts.

You can deactivate all alerts from the Classroom.

· Go to the.google.com classroom, and press Sign In.

· Sign in to your Google Account. For instance, you@yourschule.edu or you@gmail.com.

· Select Menu, at the end

· Click the Settings button. (You will need to scroll down)

· Next to Receive email alerts, pick one: To turn off alerts, click "Off." To turn on notifications, click "On."

Customize alerts

You can choose which updates you are getting for all grades. You can, for example, turn off invitation notifications for all classes, but leave notifications of assignments on.

· Go to the.google.com classroom, and press Sign In.

· Sign in to your Google Profile. For instance, you@yourschule.edu or you@gmail.com.

· Tap "Options" at the top of "List." (You may need to scroll down.)

· (Optional) Press the "On" button next to "Receiving email updates."

· Tap on any notification to turn it on or off.

For a summary of each form of notification, see the list below.

Notifications from teachers

To learn when to:	Turn on:
Someone comments on your article	Comments on your articles
Someone mentions you in an article	Comments referring you
A student sends you a private email	Private feedback on work
A student resubmits work	Student Work submissions
An instructor invites you to teach a course as a co-teacher	Invitations to co-teach classes
A planned post that was	

published or failed to post

Scheduled post published or missed

Student Notifications

To learn when:	Switch on:
Someone replies on your post	Comments on your post
You are tagged in a post or comment	Comments that mention you
A teacher sends you a private comment	Private remarks on the job
A teacher makes a task, request, or an announcement	Work and other posts from teachers.
A teacher grades or returns work	Returned work and grades from your teacher
You are invited to a new class	Invitations to join classes as a student
Your work is not submitted	

that was due within 24 hours   Due to date reminders

                                       for your work

Switch off class alerts

You can choose whether to receive updates for a specific class or not. For example, if you don't want any notifications for your Math class, you can turn them off, but for your other classes, you will still get notifications.

Note: When alerts for a class are switched off, all signals for that class are switched off.

· Go to the.google.com classroom, and press Sign In.

· Sign in to your Google Account. For instance, you@yourschule.edu or you@gmail.com.

· Tap Options at the top of the List option. (You might need to scroll down.)

· (Optional) Click the "On" button next to "Receiving email notifications."

· Click Down arrow next to "Class Updates."

· Click the" On" or "Off "icon next to the class name

February 2019

 New features

Stream notifications – if you're using the Classwork tab, select a compressed or extended view on the stream for Classwork notifications, with the option of entirely hiding them.

Flow organization — in classes that use Classwork, transfer each post to the top of the line.

Class details include necessary information, such as the class name, section and room number, Settings for posts on your stream page, your class video meeting link, and your class code. You can change class details on your Settings page.

January 2019

New features:

New look — unique style and lots of new themes.

Drag-and-drop classwork — Organize whole subjects or individual posts easily on the Classwork board.

Show and exchange a class code quickly — Class codes for each class are now at the top of the Stream tab.

Swipe for options — rapidly removes or edit posts by swiping left on your smartphone (IOS only) device.

You need to know some things before you continue using Google Classroom for the wrong reasons. It's an online learning site, but it's not:

A chatbox: You can reply to assignments and ads, but no chat feature is available. You can give them an email if you want to be in direct contact with your students, or you can allow other Google apps to take over this feature. Dream of Meet Hangouts!

A test or a quiz tool: When it comes to making quizzes in Google Classroom, there are several possibilities, but it's still not supposed to be a quiz tool. In that end, there are so many other good games. Think of quizzes about Google Forms.

Option 1: You can add assessments and assignments inside Google Classroom from other educational applications, such as an automatically graded test from some good platform.

Option 2: Inside Google Classroom itself, here's what you can do: add a question. Then select a clear answer or an item of multiple choice. It does not sound all that impressive. If you want to enhance communication in your digital classroom interactive, it's best to choose the first choice.

Discussion forum: Announcements can be made, and students can vote on them, but it's not a perfect place to talk. Check out some other featured applications if you're looking for a natural but powerful, free classroom resource that will inspire discussions (and other cool things).

Some insights into the Originality Reports and other features

Google Classroom features that are now accessible to all users were previously only available in beta. Originality Reports and Rubrics are among them.

Originality reports

The originality reports of Google Classroom act as a resource for correcting uncited content and possible plagiarism. This functionality is no longer in beta mode – it is now available in English for everyone who uses Classroom (Spanish, French, and Portuguese are in beta). To make this function applicable to an assignment, the teachers simply have to tick the box.

At present, teachers can turn on originality reports for three tasks free of charge (if the school uses G Suite Enterprise, then there are no limitations). Both teachers and students can conduct the reports at any point during the assignment. The reports include expiry dates (since web material changes continuously). Teachers will use the tool up to 3 times before the students return their homework. Teachers are allowed to access reports for each submitted paper.

If the originality report has been completed, the link to the 'view originality report' will open the story, where any problems are highlighted. The report shows the flagged content context and emphasizes the commonalities in bold. Clicking on the passage will take you straight to the questionable content website. Eventually, there will also be school-owned content repositories within each domain to test the work of the students internally. There is a toggle choice between seeing the total percentage of the flagged assignment and the number of flagged passes. The method is less about "catching" a student in misconduct, and more so about allowing them to recognize and fix possible misunderstandings before finalizing their research.

Citations using Explore

Students can use the Explore tool to cite sources that would enable students to insert footnotes in different citation formats. They simply click on the button and Explore finds connections between the topics of the documents and the content online. The paragraph in the chosen form (e.g., APA, MLA, and Chicago) is referenced by clicking on the quotation mark icon next to the correct document. Once the resource is quoted, it appears like this in a footnote: when reporting, it is expected that there is no flagged material, but this does not guarantee a plagiarism-free paper. And when passages are quoted, when an originality report is produced, the reference(s) must show up. Teachers will also use their judgment to assess if there has been plagiarism.

Rubrics

Although some briefing on rubrics has been done in the features mentioned above, it shall be discussed in some more details here. When submitting an assignment, the Rubrics feature allows students to see the grading requirements, which can help teachers grade more effectively.

The rubrics may have several parameters and point values. As of now, a numerical value must be added to each point. The details for each criterion will be shown or hidden by clicking the arrows on the right (next to the point total).

Assignment rubrics can be generated by starting from scratch or importing the requirements into a Google Sheets format. Note the time-saving tricks of duplicating criterion when constructing a rubric (click the 3-dot 'more' menu) and, of course, copying and pasting! Scoring is optional; the scores will be automatically added to the grade book in the 'Grades' or 'Marks' tab if the teachers want to rate the students' work.

Teachers may reuse a rubric from a previous assignment, or even from a different class. Open the job to be graded when you use a rubric to grade work, and press the grading button below the files button. Here you can, if appropriate, adjust the overall score and input scores for the different parameters. You may also provide private feedback for each student, just as with other assignments. Rubrics can provide timely, personalized, descriptive feedback on work for students!

Google Forms-a complementary tool with Google Classroom

Google forms are among the most valuable tools of Google Drive and are, without a doubt, one of the most reliable devices on the Internet. If you need a contact form or a checkout page, a survey, or a directory for students, away is all you need to collect the information quickly. It only takes a few minutes, with Google Forms, to make one free. Google Forms — along with Files, Cards, and Slides — is part of Google's tool kit of online applications that help you get more accomplished in your browser for free. It's fast to use and one of the easiest ways to directly transfer data to a spreadsheet, and it's the closest sidekick to the spreadsheets of Google Sheets.

Teachers may use Google forms to obtain information about the students. These forms can also be used as a sample before taking an exam. Once the students get the right question, they proceed to the next question. So if they get the question wrong, it will lead them to a support page to discuss the subject and then return to the issue to try again. Teachers will use this to collect knowledge on how they're doing for their big exams and quizzes. With several graded questions, they may build multiple-choice, short responses, essay questions, and much more. Since teachers use technology in their classrooms and often adopt the flipped model, they need to know if the students can access the internet at home and how confident they are with using technology.

It would be smart to ask their students some questions about what sort of technology they are using, how secure their internet connection is at home, how comfortable they are with using technology and what technology they are using. If the teachers have this knowledge, they will use it to help the group students and make sure they have someone good with it who will support them. They should also make sure the students who do not have the internet at home have a place to watch the videos that have been flipped.

Google Forms began life as a feature of Google Sheets in 2008, two years after the initial launch of the Sheets. A spreadsheet may be attached to a file, converted into a separate sheet, and displayed in another layer with your answers. It was necessary, but the job was done. Over time, Google added more features to the Forms, and eventually converted it into its standalone app in early 2016. Today at docs.google.com/forms, you can create and manage forms with templates and easy access to all of your ways in one place.

Google Forms is now a full-featured form application that comes to your Google account free of charge. In any sequence you want, you can add standard question styles, drag-and-drop questions, customize the form with simple picture or color themes, and gather answers in Forms or save them to a Google Sheets tablet.

The easiest way to start creating a form is via the Google Forms app. Go to docs.google.com/forms, either pick a template or open a new type. Within Docs, Sheets, and Slides, there is also a link to Google Forms: press File-> New – > Form to start a new empty form. Or, press Tools-> Create a Form in Google Sheets to start afresh, unique style that is automatically connected to that spreadsheet. This is the easiest way to get data into a new or current spreadsheet: open the worksheet where you want the content, start a form, and the answers to the form will be saved there automatically without any extra clicks.

The editor for the Forms is straight forward. Your form fills the screen center with space for a title and definition, followed by fields for the way. To edit it, select a form field, and add a question. Use the drop-down box next to the question to pick the form of a problem, such as multiple-choice, checkboxes, short answer, etc.

Google Forms provides various settings. The floating toolbar to the right helps you to add more fields to the form. You can change the color scheme of the arrangement in the top-right menu, display the way, and use the Submit button to upload the form and access other extra choices, including the installation of Forms add-ons. To see current responses to your question, switch from the Questions tab to the Responses tab in your Question Editor and add it to a List.

Google Forms contains 12 types of fields: 9 types of questions, including text, picture, and video. To add a new question, just click the + icon in the right sidebar or click on the text, image, or video icons to attach media to your form. For a simple way to add questions that are similar to your style, each field includes a Copy button to duplicate the track. There's also a delete button, options to render the appropriate area, and a menu on the right side with extra choices. You can switch the question types at any time, although note that if you transfer from multiple choice, checkbox, or menu to other question types, your field settings and questions will reset. Then, to quickly fill in field questions, just click enter to start adding another one.

Title and Description: Title and description fields are automatically added to each form and field — although the description is hidden in most areas by default — and you can use the **Tt** button to add an extra block of titles anywhere. For questions, you can leave the title and summary blank, but you have to fill in the primary type of text.

The definition does not include formatting options — although you can include links (in a condensed format, such as links.com, or as the full-length style such as https:/links.com/), and readers of the form will click those to access the site or related content.

Short Answer: This area is the best place to request tiny bits of text: names, email addresses, values, etc. To answer the question, you get one line of text — although, your users may potentially enter as much text as they wish. This area involves data validations for number, text, duration, and standard expression to ensure you get the answers you need. Number validations help you search for value ranges, while text validations are ideal for email addresses or connections to check for.

Paragraph: This is a field for text — long-form text is almost the same as the short answer field. The only data validations available here are duration and regular expression, so just use it when you want thorough input or longer notes in the answer.

Multiple Choice: The default area in a Google Form for new questions, multiple choices allow you to list options and let users select one. You can then either make the form switch to another section, based on the answer or shuffle the response options to avoid bias.

Checkboxes: Similar to multiple options, this field lists answers and allows users to select as many as they wish. It also requires validation of the data to allow users to pick a certain number of choices. This does not, however, contain segment hops.

Dropdown: Do you want all the answers in a menu? For you, this field is here then. It's the same as the area of multiple-choice — with the same segment jumping and shuffling options — only this time, the answers are in a line. It is useful if you want to keep your form compact because there are several choices to address.

Linear Scale: The area enabling people to pick a number in a linear range, allows you to set a scale from 0 or 1 to 2-10 with the lowest and highest choices labels. And yes, emoji do function for labels as well.

Multiple Option Grid: This could be the most confusing field because the fields are shown in a list rather than in the matrix when the readers see them. Essentially, you are going to add questions as rows and column choices. As many rows and columns can be created as you want, but remember that readers will have to swipe right to see more than six columns on desktop browsers or only three mobile columns. While setting up grid queries, you may want to keep the form preview open — just tap the eye icon at the top right, and refresh that page to see your changes. The grid also allows you to require a response per row, in addition to the standard response option, and can also limit users to only one response per column.

Date: Want to ask for a particular date or period, maybe scheduling an event or logging in an activity? The date field is the one you wish to pick. It may request a date and month as well as, optionally, the year and period. Remember that the date format will be shown to your position in the default format. If your Google Account is tuned to US English locale, the dates will be displayed as MM / DD / YYYY, while UK English accounts will display dates as DD / MM / YYYYY. Your users can see the date choices in the date format of your locale unless they are entered into your Google Account, so be sure to keep that in mind when designing the forms.

Time: Time allows you to request a period in hours, minutes, and (optionally) seconds to log precisely how long it took an operation.

Image: Google Forms allows you to upload a photo, insert a photo from a connection or Google Drive, or take a picture from your webcam (as long as Flash is installed). Or, Google Photos can be searched for images, including royalty-free stock pictures and Life photos that are allowed to be used inside Google Drive.

Video: Google Forms support YouTube videos only, which you can find by searching or by connecting to a link.

If you have incorporated photos or videos, your entry form will be compatible with the standard title and description, along with options for resizing and displaying the video or picture-oriented, left, or right.

Form Sections and Reasoning

Simple communication forms require only a few fields, but longer surveys on one page can easily get daunting with hundreds of questions. That's where sections come in handy: to answer one set of questions at a time, they let you break up your form into parts. Just click on the right toolbar at the last button to add a section below the current issue. Every section contains its title and definition, as well as an arrow button at the top to view or hide questions and to keep your form editor clean.

Although you can drag-and-drop questions between lines, you cannot rearrange full lines. Instead, you might push out the questions and then delete the line. And, if you wish to repeat a segment, simply click on the menu of the segment and pick Duplicate section for another copy of those queries. That's the perfect way to start a logic-jumping form. Say you'd like to ask a respondent follow-up questions based on their answers — maybe ask which meat a participant needs, but only if they're not vegetarian.

Simply add sections with the optional questions, then either add a section jump to the multiple selections, checkbox, or menu questions, or the section itself. Make sure you think about where people who shouldn't see these questions are also sent, maybe in a different section with alternative questions. Or, if there is nothing else to ask, you can give them straight up to the end of the form to request their responses.

Be creative: Form sections and jumps allow you to turn your form into a mini-app, and they can be a great way for each person to condense comprehensive surveys into only the most relevant questions.

Create a Quiz

Another way to do an interactive form is via the Quiz feature of Google Forms. You can find a tab of Quizzes within your settings for the form. Select Make this a questionnaire, and then select whether to show the results immediately after submission of the form or later after reviewing the answers. If you choose the latter, the form will demand that respondents sign in with their Google account. If you choose, you can then choose to display the missing and right answers, as well as a value for each choice. You can see a new Response Key button at the bottom left of each question, with this activated. Tap on it, then pick the right answer to the question. Optionally, you can provide comment reviews for both correct and incorrect responses, with a link for respondents to see more details if you wish.

Plan Your Form

There's one place where you have no option: your form's design. Google forms include a color or image header, and a lighter color accent as the background. New forms come in purple by default, while the prototype shapes also have an image. To tweak your style, press the color palette icon at the top-right if only a little. You can choose from 15 colors with a complimentary background hue, each one a darker color to the header.

To pick a photo or Google Doodle-style drawing from Google's library, click the photo icon as the header photo of your page. Or, pick one of your Google Drive photographs or upload a new one and crop it to fit in as a header for the form.

Forms will then automatically pick a color matching your picture history.

Some of the header photos used are animated GIFs featuring burning candles, rolling balls, and more. Sadly they appear as a regular still image if you add them to your form. Google Forms may be getting support for GIF in the future — for now, icons and colors are the only design choices in Forms.

Store Form Responses in a Spreadsheet

Once the form has been generated, you do not need to do anything extra to store answers from the respondents in Google Forms. By default, it will save any answer in the Responses tab, showing overview graphs and answer lists. A view of an individual response shows the live form along with each respondent's results. That's great for quick form responses, but you can connect your form to a Google Sheets spreadsheet for more tools to analyze answers. Just click in the Responses tab on the green Sheet icon or press. Select Answer Destination in the menu, then create a new spreadsheet or pick an existing one to store the answers.

One great thing about saving entries from Google Forms to a spreadsheet on Google Sheets is that it's easy. Change the field names of your form, and they will be changed automatically in your table document. Get a new entry, and as soon as your recipient clicks Submit, it will appear in the list.

Google Forms still maintains a complete copy of all data about your form, so don't worry if you accidentally delete anything from your spreadsheet. Just open the answer settings for your Form and unlink it from your spreadsheet, or press Form-> Unlink Form within your spreadsheet. Then reattach the form to your spreadsheet, and Google Forms will restore all the data from the form to a new file.

Share completed forms online

Ready to receive responses? Select the Send button to share the form via email or social networks in the top right corner, copy a connection to the form, or get an embed code to add it to your site. You can either copy a full-length link with the link or get a shortened goo.gl/forms/link to share on social networks more easily. The embed choice provides options for the width and height to suit the shape within the context of your web. Sharing the form via e-mail requires an additional option: sharing the e-mail form. This copies into the email your actual form choices, and if your user uses Gmail, they can fill in the form in their Gmail inbox, press Send, and send in their reply without ever seeing your original form. It only works in Gmail, though — Apple Mail displays the fields of the form but doesn't submit the answers to Google Forms, and Outlook.com can't even open the form — so you may want to attach a notice to non-Gmail users with the request.

Share Paper or PDF Form

Need to collect offline responses? Here too, Google Forms will help. Just press Print in your Form tab, and a ballot-style copy of your form will be made by Google Forms that you can print or save as PDF.

Grids and multiple-choice options display pill buttons to fill in, while text fields contain clear answers. Simply type your answers into your Google Sheets spreadsheet to save them along with your other entries in the form until the respondents have filled out your paper forms.

The Google Forms are frequently used by the teachers when operating the online classrooms. The quizzes, assignments, and other tests created using these forms would help ensure the effectiveness of Google Classrooms.

How is Google Classroom supportive of classroom differentiation?

Google Classroom may help streamline the formative assessment, which is essential to help students who may need more support or additional questions. For instance, you can use the platform to build, distribute, and collect digital exit tickets or auto-graded appraisals quickly. Google Classroom can, in a way, make it easier and faster to gather daily feedback on the progress of your students. There are, of course, plenty of other formative evaluation resources out there, many of which now provide Google Classroom integrations.

Google Classroom also makes it easy for individual students or small groups to customize assignments. This means teachers will give other students or classes in a class changed or different tasks. You also have the option to check-in privately with a student to see if they have questions or need some extra help. The ability to do all of this online may make the distinction efforts of teachers less visible for the class, something that could be beneficial to students who might feel singled out. Differentiation will always be a matter of creative problem-solving with or without a tool like Google Classroom, and there is no one or "right" way to do that. Fortunately, many teachers post online their ideas, strategies, and innovative solutions.

3.3 Simple steps for setting up Google Classrooms

Google Classroom's simple setup method is relatively straightforward, even for first-time users. The Google Teacher Centre offers multiple tutorials to get started — if you are looking for the most popular videos and details, this is your best bet. There are also plenty of do-it-yourself tutorials shared by professors and software development experts on YouTube. Some of these videos produced by teachers provide practical tips and techniques they have learned in their classrooms by using the site. Okay, so, now that you've come this far, that means Google Classroom has to say something to you! You would find it easy to set up and very intuitive to continue to use it. Follow these steps to set up your teacher account at Google Classroom:

Step# 1: Signing Up

Since we speak to students about using Google Classroom, it is presumed that you are in a school or district that uses G Suite for Education. It's also believed you know your Google login information that your school or IT department has given to you. Finally, it is thought that you must log into the Classroom from a laptop or device connected to the internet. Before trying it on a smartphone or tablet, it is recommended to do so. You can use the Classroom by logging in with a G suite e-mail address when you go to classroom.google.com, or you can use it for educational purposes without claiming to do so. This way, it works just fine too. If you have hundreds of pupils in your class, it's just harder to manage your students. You're going to have to add them one by one.

Step# 2: Set up a class

As a teacher, designing a class is one of the first things you'll do in the Classroom. In a class, students will be assigned work and get to have announcements posted by the teachers. If you are teaching multiple classes (at the secondary level), then for each section you are teaching, you would create one class.

Adopt the steps below to create your first class:

- Go to classroom.google.com and log in.

- Select the role of the teacher.

- Click the + icon on the home page of the Classroom and then pick Create class.

- Give the class a title that makes sense to you and your students.

The following are optional but might be relevant to your teaching situation:

· Click Section and enter the details to enter a short description of your class, grade level, or class time.

·Select Subject to add a subject such as Algebra I, and enter a name, or select one from the category that appears when you enter text.

·Click Room to enter the classroom venue and enter the info.

·Click on the Create button.

You can now see a class code shown, but that's not going to be necessary right away. When you can invite students to your class, you will come back to that at a later date. If you need to see the code at any time, you can display it on the Stream tab.

Congratulations, you now have your first Google Classroom created. You're well on your way to improving student learning by using Google Classroom.

Step# 3: Inviting students to the Google Classroom

Once you've built your class, you can invite your students to participate. Let them sign up by entering the unique code you've given them using the Google Classroom app. You can find the code in your class which was developed. Go to the "students" page. Another choice is to allow your students to enter their e-mail address, one by one. One thing you should keep in mind is that your students need an e-mail address from Gmail or Google.

You can also have your students visit classroom.google.com to let them in. You can select "join class" there, enter the class code, and your students are in! This could be a little faster as you don't have to type in the e-mail address of every student. Now get ready for your online lesson! At least, it's there, and it's accessible to everyone. You have to do a few other things before you can take off for good.

· **Make your first task, or an announcement:** You can share a new statement in the Stream, or go to Classwork-click on the "+ Create" button and share your first assignment to Google Classroom. Don't forget to have your tasks numbered. The students will find it easier to see which one comes first because you can't reorder tasks in the stream. You can also transfer tasks up to the top. Press the title to see if any students have handed in the assignments. Also, give grades and feedback later on. You will then return the tasks to your students so that they can start editing again.

- **Attach some lesson material to your curriculum/task:** Fill in Google Drive material or add a YouTube video, a computer file, a connection, etc. You will find those options right below the due date. If you just want to share your class presentation, which is not related to an assignment, you can go to the "About" tab. A few lesson materials like slides, interesting papers, and examples can be added here.

- **Open the folder on Drive:** Each time a new class is created, Google Classroom creates a Drive folder for that class. You can navigate the folder by going to all tiles in your lesson. You can find a folder icon on each piece of tile. Click on it, and you will be in the folder. You can add materials for the class here too. Most of your student assignments end up in the Google Drive folder automatically, and you'll get it back whenever you want.

3.4 Adding and Grading Assignments

Although this feature has already been highlighted to some extent in the discussions above, this section will bring in some in-depth info on adding and grading assignments using Google Classroom.

Making an assignment

Students just love homework, and as a teacher with Google Classroom, you can easily add assignments that include the attached content. After the assignment is posted, the students receive an e-mail notification of the task, after which they complete it and return it. The cool thing is that they lose edit rights for that assignment after the students turn in the assignment, which means they can't change it. Then, the task can be marked.

Here's how to make your class assignment:

1. Log in to your class and click on the Stream tab if it's not already displayed.

2. Select File.

3. Type in the assignment title and optional definition. The description is a perfect place to bring the task directions in.

4. If you need to, press the due date to change it.

5. Press "Add Time" to add the time of day when the assignment is due.

6. Click the correct icon:

· **Paperclip:** This option adds a file to the assignment if you have materials to add to the assignment. You can import the file from your hard drive here, which is saved in your Google Account.

· **Google Drive:** This option always adds a file to the task, but lets you locate the data right on your Google Drive.

· **YouTube play button:** This option allows you to add to the assignment a YouTube video. You can either check for the video on YouTube or copy or paste the YouTube video URL when you press this button. YouTube results are shown right in the same window while you are looking for a video, and you can even preview the video, so there is no need to visit the YouTube site.

· **Link:** You can press the chain-link button on the chain to paste the assignment in an external URL.

You can grant permissions for documents that you upload or select from Google Drive for what the students will do. To allow students to do the following, click the drop-down list:

· **View only:** Select this option if you want all of your students to read the same data but don't want to change it. That is only good for materials of comparison.

· **Edit:** If you want all of your students to make changes to the same file, choose this option. That is only good if students are supposed to work together on a single assignment.

· **Make a copy for each student:** If you want each student to have their copy of the assignment, choose this option. Students will make adjustments and make different turns in the task. This is ideally applied to traditional homework assignments where the student is responsible for their job.

7. Click Assign.

The assignment is made, and every student is informed of the task by e-mail. The task appears on the Stream page of the class, on which you can check how many students have completed the task.

Although you can upload files created in non-Google applications, such as Microsoft Word, it is easier to use the forms of Google Drive to build documents if you are going to share them in this way. Google Files, Sheets, and so on are entirely incorporated into the Classroom, and when you complete your assignments, your students won't have to jump through the hoops. For instance, if you upload a Microsoft Word document, the student would have to download the document to complete it, re-upload it when it is finished, and add it to the assignment again. Or they would need to open the file in Google Docs to complete it and re-attach it to the assignment. It removes most of those steps by building the document in Google Docs, to begin with.

Learners and students will report on the assignment at the bottom of the assignment on the Stream tab. Here you can provide additional explanations, or students can give general feedback about the assignment or ask questions that all the other students will see.

Grading the assignment

With Google Classroom, the days of learners handing out papers and homework assignments are gone. As an instructor, you can make your assignments with any extra resources, such as handouts or worksheets, in your curriculum. Students complete and return the assignment-all electronically on Google Drive. You can also rate them online after your students turn back into their assignments. Here's how:

1. Sign in to your class and click on the Stream tab if it doesn't already show up. In the center column, you will see the assignments that you have made.

2. You can see how many students have completed the assignment in the Task box, and those who have not completed it. Click on the number above the "Done" option. You can then see the list of students handing in the task.

3. Select the student's name to extend his / her assignment.

4. Tap on the folder attached to the student's assignment to see what the student turned in. The document opens in the appropriate Mobile app (e.g., Mobile Docs).

5. Make any remarks inside the document that you have.

To stand out from the student's text, you should type your comments differently. Much as in the old days, when the teachers marked a homework assignment using red pens, you can use red text to add feedback. Or, better yet, the input can be received using the Comment feature. Simply highlight the text you are commenting on, and select Insert and then comment option. Type the comment in and press Comment. All of your comments are stored automatically in the document the student turned in.

1. Close the document to go back to the work page for the students.

2. Select where it says No Grade to grade the work.

3. Type the number of points awarded, between 0 and 100. In this area, letter grades are not recognized.

4. Check the box beside the assignment for the student.

5. Select "Return." Assignments have to be returned to the students before documenting them.

6. The classroom will ask you if you still want the assignment to be returned, and if you want any input. Upon completion, click Return Assignment.

7. The task shows in the assignment list as returned. The student receives an e-mail that you returned the assignment and, if necessary, can edit the assignment, and return it.

When making assignment worksheets, it's best to use the apps from Google Drive, such as Google Docs, Papers, etc. This is because the Classroom is wholly incorporated into Google applications. If you are using third-party programs, such as Microsoft Word, then you and your students need to download the files, re-upload the data, and re-attach them to the task. Using Google apps does away with all the hard work.

Pro time-saving tactics while grading assignments

Some teachers add a private message, for each student, in each

task when grading in Google Classroom. And while on some tasks, each student needs unique comments, others allow the teachers to repeat comments.

The comment bank in the grading tool for Google Classroom is super helpful, but it is often found that only 3-4 clicks can often add up to insert a comment, mainly while working with several students. That leads to the invention of a pro tip for Google Classroom time-saving: submit a private message to multiple students at once. Here's the flow for this:

· Teachers, with the Classroom's grading tool, can go through the assignments of the students, one by one. If the student needs a specific comment, then they can grade, comment, and return from the assignment tool area the work of that student.

· Teachers can only enter their grade on the assignment tab, and no feedback, for the students who meet the criteria for an assignment.

· If the teachers have finished all the student work, and are returning all the assignments with specific feedback, they can go back to the Assignment Student Work page. This is the page where they can see all the tasks of the students in one location. There is a list of who's turned in, who's still missing, and who's been graded. It is here that the time-saving magic takes place!

· Select all of the turned-in assignments with a single click, and then hit "Return." At the bottom of this pop-up box is a private comment option. Type the positive feedback, then click Return, and Google repeats this private comment for every student and returns all assignments in one click!

Why not free up as much as you can to spend on what you love? In education, opening up an entire day, afternoon, or even an hour can be incredibly difficult. Professionals must be time scavengers, demanding minutes along the way, and those minutes will add up to hours if they are intentional.

3.5 Privacy Evaluation for Google Classroom

The terms of Google state that they use knowledge to help improve the services' protection and reliability. They state that they make contractual obligations in their agreement on the G Suite for Education and that they are committed to compliance with the privacy and security requirements.

Safety

Google uses knowledge to help improve the services' security and reliability. This involves identifying, preventing, and reacting to fraud, harassment, security threats, and technical issues that could affect Google, its users, or the public. A school will offer students access to Google resources like

Google Docs., Sheets, Slides, and Sites. These aids allow students to interact in real-time with their peers and instructors, allowing them to share their work, gain feedback, and make edits instantly. They may be kept private, shared, or even made public, with others (such as a parent or the entire class. When users share information publicly, search engines, including Google, can index the information. The services offer different content sharing and deletion options for users.

Privacy

Privacy G Education Suite allows users to create a Google Account, which is developed and maintained for students and educators to be used by a school. The terms state the school may provide Google with some personal details about its students and educators when establishing this account. In most cases, it includes a user's name, email address, and password but may also include secondary email, telephone, and address if the school wishes to provide that details. Google may also specifically collect personal information from users of G Suite for Education accounts, such as telephone numbers, profile photographs, or other information that they attach to a G Suite for Education account.

Core features of the G Suite for Education include Gmail, Calendar, Classroom, Contacts, Drive, and Docs. Sheets, Slides, Sites, Talk / Hangouts, and Chrome Sync. Under its G Suite for Education agreement, these facilities are made available to a student. In addition to the Core Services, users of the G Suite for Education will have access to other Google services commonly accessible to customers, such as Google Maps, Blogger, and YouTube. Such concepts are considered "additional services" because they are beyond the core programs of the G Suite for Education.

For G Suite for Education users in primary and secondary schools (K-12), Google does not collect or use any personally identifiable user information (or any information relevant to a G Suite for Education Account) for advertising purposes. No data is taken to create advertisement profiles, whether in core services or other Google services accessed while using a G Suite for Education account. Parents and educators, however, should be aware that Google may serve advertisements in the "additional services" to G Suite for Education users. Still, administrators have the right to limit access to those additional services. Finally, the terms of Google state that they do not own any user data in the core services of the G Suite and do not distribute or sell G Suite data to third parties.

Security

Google Security terms state that they are entirely committed to the protection and privacy of user data and that they protect users and schools from attempts to misuse it. Google claims that its systems are among the most reliable in the industry, and oppose aggressively any unauthorized attempt to access data from customers. The terms of Google state that all facilities used to store and process user data comply with fair safety requirements that are no less stringent than those in facilities where Google stores and processes its similar type of information.

Google's terms further define that it has introduced industry-standard systems and procedures to ensure the protection and confidentiality of user data, to protect against potential threats or hazards to user data security or privacy, and to protect against unauthorized access to or use of user information.

Furthermore, Google's data centers are using new hardware that runs a new rugged operating system and file system. For protection and efficiency, every one of these systems has been optimized. The terms state that since Google manages the entire hardware stack, it can respond rapidly to any threats or vulnerabilities that might arise. Google's terms and conditions say that it must take reasonable steps to ensure that its staff, contractors, and sub-processors comply with all security requirements to the degree relevant to their scope of operation. These steps include ensuring that all individuals allowed to process personal data have committed to confidentiality or are subject to a reasonable statutory confidentiality requirement.

Google's terms also state that they encrypt Gmail and Google Drive data (including attachments). Additionally, user data uploaded or generated in G Suite services will be encrypted at rest. The terms state data are encrypted into multiple layers. Google forces HTTPS (Hypertext Transfer Protocol Secure) for all user-to-G Suite transmissions and uses Perfect Forward Secrecy (PFS) for all of its services. Google also encrypts message transmissions using 256-bit Transport Layer Protection (TLS) with other mail servers and uses 2048 RSA encryption keys for authentication and key exchange levels. This prevents the exchange of messages as users often use TLS to send and receive e-mails with third parties. PFS does not allow private keys for connection to be stored in permanent storage. Anyone who loses a single key will no longer decrypt connections worth months; also, HTTPS sessions are not retroactively decrypted even by the server operator.

Finally, if Google becomes aware of an unauthorized data infringement, the terms state Google would immediately and without further delay inform users of the data infringement, and immediately take appropriate action to mitigate harm and protect user data.

Compliance

Google's terms of service state that they make contractual obligations in their G Suite for Education arrangements and that they are committed to complying with privacy and security requirements. Whether it's real-time dashboards to check system performance, ongoing Google process auditing, or sharing Google's data center location, the terms state that Google is committed to providing full accountability to all its users.

The terms of G Suite for Education state that its core services comply with the Family Educational Rights and Privacy Act (FERPA). Where user data contains FERPA Education Information, Google will be deemed a "School Officer" (as defined in FERPA and its implementing regulations) and must comply with FERPA. Additionally, if schools allow users under the age of 13 to use G Suite for Education, Google's terms state that they contractually require schools to obtain parental consent using G Suite for Education as specified by COPPA. Schools are also expected to obtain parental consent for the collection and use of personal information in the "additional items" that the school may choose to use with students before allowing any End User under the age of 18 to use these services.

In primary/secondary (K-12) schools, parents of G Suite for Education users may access their child's personal information, export the data or request that it be removed, via the school administrator. School administrators should have parental control, export, and deletion of personal information that is compatible with the services' functionality. The terms further state that if a parent wishes to avoid any further collection or usage of the child's details, the parent can request that the administrator use the service controls at their disposal to restrict the child's access to features or services, or remove the child's account entirely.

3.6 Apps for creating content for Google Classroom

Google Classroom coordinates with hundreds of educational programs. These integrations save time for teachers and students and make sharing knowledge between Google Classroom and their favorite apps seamless. Below is the list of some useful applications which work closely with Google Classroom to make learning an enjoyable experience.

Actively learning

This constructive learning application works smoothly with Google Classroom. Teachers can quickly synchronize Classroom rosters to actively learn and synchronize Actively learning assignments and grades back to Google Classroom.

Additio App

Additio App is a collaboration package that lets teachers stay organized and contact students and families efficiently. It provides other useful tools, such as a strong grade book and a durable lesson planner.

Aeries

Teachers can connect or build new classes based on their Aeries classes, and import scores into the Aeries grade book.

Aladdin

This integration enables the automatic development of Google Classroom classes based on classes in Aladdin. Assignments and grades between Aladdin and Classroom can also be synchronized.

Alma

This app is the first Student Information System to provide full integration with Google Classroom. Teachers can synchronize assignments and grades with this integration, and tech teams can deal with the creation and management of Google Classroom classes through their schools and districts.

American Museum of Natural History

The American Museum of Natural History provides Educational K-12 services and resources. To share related papers, curriculums, and tools, use the Share to Classroom option.

Aristotle Insight: K12

This all-in-one classroom management, content filtering, and monitoring system empower students to become informed, and safe digital citizens.

Assistants

Provides synchronous feedback to teachers and students when students complete assignments using this free online resource.

Book Widgets

Book Widgets provides collaborative workout models. To engage students, teachers can choose between over 40 different widgets or templates.

Brain POP

With Brain POP, teachers can import their classes directly into My Brain POP from Google Classroom. SSO-ready student accounts are created when a teacher imports a class, enabling students to log in to Brain POP through the Google launcher menu.

Buncee

A design and presentation platform to develop immersive educational material for students and educators allow learners of all ages to imagine ideas and connect creatively. Simply build your task, note, class reminder, activity, or project and share it in your Google Classroom with the students.

CK-12

CK-12 platform provides a library of free online textbooks, animations, quizzes, flashcards, and real-world applications for over 5000 topics ranging from arithmetics to history.

Classcraft

With the incorporation of Classcraft, teachers will take a single click to pull rosters from Google Classroom and provision accounts. Teachers could give timely turning in points to the students' assignments submitted well in time, in the game, and translate their Classroom results into points of the game.

CodeHS

CodeHS is a robust framework designed to help schools teach informatics. They have online instructor tools and resources, and professional development curriculum.

Curiosity.com

Their mission is to spark curiosity and encourage learners. This app develops and curates engaging topics each day for millions of lifelong learners around the world.

Nearpod

Nearpod is a presentation device. It is a lot more than that! Make your immersive introductions. Add some slides, slide by slide or select a particular Sway template that you can adjust. All of those slides make an excellent interactive presentation. Especially when you're introducing activities such as quizzes, open-ended questions, surveys, questions are drawing, and others. Inside your presentation, what about taking your students on a field trip? Only add a slide from Nearpod's library, featuring a virtual reality experience.

When your presentation is ready, your students can choose to enter a code in their Nearpod app or just click on the connection in Google Classroom assigned to them. You are in charge of the interpretation as an instructor. If you turn to another slide, the students' presentation on their devices will turn to that slide as well.

If your students have to do a quiz or questionnaire, they could do it on their computer, as it is part of the presentation-a live set of responses! So, you can see how your students responded immediately.

Duolingo

Is teaching a language a hard job to do, and a hard thing to understand for students? Duolingo is also one of the most popular Google Classroom applications to use! This app works well and is seen as the most popular language teaching and learning device in the world. Through the 23 languages it provides from Spanish to French, Russian, Dutch, Swedish, Italian, Hebrew, and more, your students can learn to be fluent in any language by continually practicing with this program.

Duolingo teaches a language through enjoyable lessons of a bite-size nature. It also helps you to record yourself talking and seeing what speaking in another language is like. Although Duo the owl keeps track of how well you're doing, you can practice talking to Bots in real terms! And yes, Duolingo is linked to Google Classroom, and it's one of the best applications to use for teachers.

Chapter 4: Google Classroom-An Interactive Platform

If you want to build a more connected platform for students, you might be considering doing so on the Stream page of Google Classroom. The Stream is a feed within Google Classroom where everybody in the class can find announcements and upcoming assignments, and it is the first thing students see when they log in.

Some teachers use the Stream to set up class discussion boards, where students can connect online by asking questions or commenting on the posts. Such discussion boards will help improve class engagement and give more leverage to students in getting their voices heard (or read) by the teacher. You can use the Stream as a closed social network of sorts with conversations, and it can be a great way to help children practice using all kinds of different digital citizenship skills in a "walled garden" style environment. Google promotes interaction in its classroom application to ensure a more significant outcome.

4.1 Engagement through student-teacher interaction

Once a student logs in to complete an assignment, they will make a class comment to which all other class-fellows and the teachers assigned to that class will receive a notification (through email and app notification) that students and teachers will reply to. This can be an enormous opportunity for both the teacher and students because they can answer a question or assist the whole class with a misunderstanding.

A student can also send a private message to the teacher if they want to ask a question without their classmates' prying eyes. After all, the same social problems are evident in the digital world as in the Classroom (How many times did you, as a teacher, have to deal with student social media conflicts?).

The following ways can bring in more significant student-teacher interaction with Google Classrooms.

With Google Classrooms, teachers can:

· Organize, distribute, and compile assignments, materials for the course, and student research online. Teachers are often able to post a task to different classes or to change and repeat assignments year after year. This would bring in some interaction between teachers and their students.

· Communicate about the classwork with the students. They can use the site to post announcements and notes about tasks, and it is easy to see who has finished their job or who has not. They may also check in privately with individual students, answer their questions and give support, as already mentioned in the beginning.

· Offer timely feedback to the students on their assignments and assessments. Google Forms can be used inside Google Classroom to build and exchange quizzes, which are automatically graded as students turn them in. Not only would the teachers spend less time grading, but their students will provide direct feedback on their work.

Engagement through videos

Educators can provide grades or reviews electronically without ever having to deal with paperwork by using open technology. Additionally, all work on the course is saved so that students can revisit it while they are at home. Students can also complete assignments via Google Classroom and communicate with teachers. This two-way communication method makes teaching and learning using the platform more convenient. By incorporating video, it makes engaging students even more comfortable.

Below are some reasons why videos could be helpful in a Google Classroom:

· **Video facilitates collaboration and learning:** In Google Classroom, multimedia is used by educators to improve the course work. Many are making videos within their class as interactive learning resources. By using video platforms, educators can create video tutorials or lessons, provide student input, use as assignments for students, or capture lectures all with a click of the record button.

· **Feasibility in access:** By using videos, educators can interact effectively and keep students learning without ever having to waste time in class. The videos are sent home and viewed in flipped or mixed learning scenarios. The student will learn from home, which makes them more interested in the Classroom.

The Classroom is an online application that can be used anywhere. Educators are given access to their videos on multiple devices with an account. They can quickly jump between devices and have access to video recordings.

With the Chromebook apps, educators and students both can record and share their videos. Videos can be stored directly on Google Drive. All files uploaded are stored in a folder in the

Classroom. This makes videos easily accessible to teachers as well as students.

- **Saves Time:** Video is incredible for time-saving. Forget about typing out long assignments or grading documents! With video, educators can film assignments and be able to assign them all in a few minutes. The educators will add a video file with instructions when they make an assignment in a Classroom.

- **Encourages teamwork and communication:** Video encourages collaboration and strengthens conversations. Google Classroom gives students several ways to work together. Teachers can encourage online student-to-student discussions and create group projects within the tool. Students will hold talks with each other with video and complete tasks as assigned to them. Also, students can collaborate on Google Docs and share their work with teachers easily.

It is an immersive learning environment that is collaborative. By using videos, they can further enrich the experience. With Classroom, teachers can separate assignments, integrate videos and web pages into classes, and create shared group assignments for students.

- **Strengthens the student-teacher bond:** Video provides a more reliable link with the students. Positive feedback is needed for students to learn. This is a worthy aspect of all learning. So why not do it by video? Recent studies have shown that at a higher level, video mentoring and feedback requires students to communicate with teachers. It gives them a bond that they would otherwise not get in a group environment. Video offers a one-to-one friendship without being face to face.

Google Classroom educators can easily grade assignments. They can give any student personalized feedback. There's also the opportunity to comment on the grading tool. Additionally, the Classroom app in smartphones helps users to annotate research. Google Classroom can save all kinds of grades quickly.

Record the video, go to the screen recorder and press the 'record' button. You need to upload and publish the video when it is finished and provide the students with the link. They can access it from anywhere.

Film and share every instructional video with your students. You can monitor them while they watch the video lecture. You will be able to see if your pupils watched your video. You will also realize when the students started watching exactly, and to what section they kept coming again. Video analytics help you understand what interests or engages your students, which part of the video needs further detail, and where your students lose interest.

4.2 Engagement through student-student interaction

Originally, while using the class comment feature initially, teachers did find some sorts of distractions-there was all the usual chatter typical in social media design. However, once the students started using Google Classroom, the teachers began to note a slightly unexpected advantage of the class comment feature. Students began to answer each other's questions. In their online Google classrooms, not all classes or students do this, but the ones who do excel. While teachers need to step in and answer a few questions, students do teach one another for the most part!

The SRS (Student Response System) built into the platform is a prominent new feature. This helps teachers to inject questions into the stream page of the Classroom and start question-driven discussions with students answering each other's answers. Teachers may post a video, photo, or article, for example, and include a question that they want their students to answer. This way, teachers can learn and check in on the progress of their students, which is a fundamental practice. They can do that very quickly with this new functionality, from anywhere at any moment.

To increase interaction among online students, teachers can assign them group projects. Forcing students to work together will add new experiences for the students and contribute to strong collaboration among them. The most efficient way of learning is group learning. This offers students a chance to support their fellow mates and to learn to work together. Teachers should get the students together in small groups to prepare and let them and their team create a video project. They may ask them to take photos, record meetings, and upload and complete the project documents such as pictures or audio files.

However, if the students don't get along or their work styles aren't compatible, it can also backfire. Online, this dynamic can be exacerbated because students work with only a limited understanding of the personalities and activities of their fellow students.

4.3 Parental inclusion in Google Classroom

With the Google Classroom App, you can connect parents' email addresses to their kids, which helps parents to monitor their child's home learning closely. Parents, although, cannot see or engage with class feedback, they simply receive an email notification that their child has a home learning assignment, so it is important to have parental buy-in to ensure that students develop and accomplish as much as possible. Google refers to parents and families as "guardians" who may elect to receive summaries of unfinished assignments, upcoming assignments, and other class activity by email.

One way to ensure greater parental involvement in Google Classroom learning experience is to organize the evening meetings for the parents and teachers. Teachers and staff can use Google Classroom as a centralized place to book evening appointments for parent-teacher consultation. All instructors are included in Google Classroom, and they can consider the creation of an appointment sheet much as and when students are given an assignment. In such a form, teachers can book meeting dates with their students, and the school administration would immediately know when the appointment is. This might help make the whole evening coordinated much better and run smoother.

Chapter 5: Google Classrooms-Advantages and Limitations

The Google Classroom analysis below states some benefits and drawbacks to help you determine if Google Classroom is suitable for your e-learning courses.

5.1 Advantages of Google Classroom

1. Simple to use and easy to access from any device

Even if you're not a regular user of Google, using Google Classroom is one piece of cake. Aside from being provided via the Chrome browser, it can also be used from all laptops, cell phones, and tablets. Teachers may add as many learners as they want. They can create Google Documents to handle assignments and updates, upload YouTube videos, add links, or download files from Google Drive very easily. It will be equally simple for learners to log in, as well as to collect and turn in assignments.

2. Good connectivity and exchange

One of Google Classroom's most significant benefits is Google Docs. These documents are stored online and shared with an infinite number of people, and when you make an announcement or assignment using a Google Doc, your learners can access it directly through their Google Drive, as long as you share it with them. Besides, in Google Drive files, Google Docs are conveniently stored and customized. In other words, to exchange information, you no longer need emails;

you just build a text, exchange it with as many learners as you want and voila!

3. The assignment cycle speeds up

How about making and distributing an assignment with just a click of a button? And how about learners turning in a matter of seconds their completed tasks? Making and turning in assignments has never been quicker and more effective, because you, as a teacher, can quickly check in Google Classroom who submitted their homework and who is still working on it, and give your feedback instantly.

4. Proper feedback

When it comes to feedback, Google Classroom allows you to offer your online support to your learners quickly. This ensures feedback becomes more prosperous as new reviews and remarks have a more significant impact on the minds of the learners. Google Classroom is a resource designed to help teaching and learning. It is an excellent interactive forum for the students along a course or level that the instructor can customize according to their teaching style and community profile and objectives. There are several ways for teachers to use the Classroom. In essence, it can be used in a teacher-led conventional way or used more creatively in a more contemporary manner, which in effect will lead to more innovation and collaboration among students. There is a diverse range of content available online — websites, classes, YouTube videos, forums, etc. — that provides tips, strategies, and creative ideas to help teachers use Classroom in creative and inventive ways to match their learner needs.

5. No paper required

There might be a day when it would be impossible to imagine grading papers; Google Classroom is undoubtedly keen to get there as soon as possible.

You have the opportunity to go paperless and avoid thinking about printing, distributing, or even losing the work of your learners by centralizing e-Learning materials in one cloud-based venue! Google Classroom allows students' access to resources, no matter where they are since everything is posted online. Students cannot lose their research in case they have physically lost it in their presence. Since they typically operate on Google Drive, everything is immediately saved, and excuses are dwindling. Students will encounter more organizational performance with a few short lessons about how to use these online resources properly. Hence, gone are the days of lost worksheets or rubrics. When required, absent students can easily access classroom resources from home – this will also help save teachers and their students a lot of stress in the long term.

6. Clean interface, and user-friendly

Google Classroom invites you while remaining loyal to clean Google layout standards, in an atmosphere where even minute detail about the design is simple, intuitive, and user-friendly. It goes undoubtedly with the saying that users at Google will feel right at home.

7. A good device to comment on

For a variety of online courses, the learners may comment on specific locations inside images. Teachers can also create URLs for new comments and use them for further discussion online.

This has been shown frequently that technology engages the students. Google Classroom can help students get involved in the learning process and remain active. For example, if teachers have students answering questions in the Classroom, other students will comment on those answers and expand thought for both students.

8. It is for everyone

Educators can also enter Google Classroom as learners, which means Google Classroom can be set up for you and your co-teachers. You may use it for faculty meetings, exchanging knowledge, or professional development.

9. Language and competencies

When the teacher who produces the class or group of students shares content, teachers may take charge of the language levels and keep them related to their learner community, using language at all levels as appropriate. Teachers can slowly develop the learning environment and distribute the course materials at the speed of their classes, depending on the subject requirements and group profile. The choice of which students they are inviting to specific classes enables teachers to delegate work based on the particular learning needs of their students. Teachers can build classes of up to 1000 students and 20 teachers, allowing teaching by a team where appropriate.

10. Content of language learning

A handy feature of Google classroom is that once they've submitted it, it helps students to go through their work. Teachers will get updates about the students' reworks or feedback on something they find difficult. This ensures they can give the students who need it, individual attention, and give them more opportunities to show their learning, operating at the right speed for them. Teachers can easily discern knowledge by determining which students may need extra help, who might want to work with response grids of model responses, etc. The Google Translate plugin for the Classroom is also available to English language teachers.

11. Exposure to an online world

A lot of colleges today expect students to take at least one online class during their degree research. If one gets a Master's degree in education, some of their online coursework might be eligible. Sadly, many of the students never had any online education experience. That's why, at a young age, teachers really should make sure that their students have as much exposure to the online world as possible. Google Classroom is a simple way for students to assist with this change because it's super user-friendly, making it a perfect technology intro.

12. Differentiation

Google Classroom is an ideal resource for differentiation, as teachers can set up several different classrooms. If teachers focus on a topic in the Classroom and have groups that focus on two different levels, they can simply build two different classes for that subject. This means they can reach out to those who struggle with their kind of job without making them feel bad or dumb.

This can help teachers offer assignments on a more individual basis, and can also really reach out to some students. They can even break people into groups where teachers think they can work the best together. Google Classroom is a perfect, versatile way to make sure every student gets what they need, and as instructors see fit, they can quickly delete and recreate classes.

13. Saves time and cost

Students lose out on all of the 'hidden' costs of studying at an institution by taking online courses with Google Classroom. This includes travel costs (which in some cases are very high), the costs of printing out assignments, and so on, and the stationary and notebook costs.

Although it is difficult to determine how high these costs would be before the students enroll in a course, some important considerations should be pondered over. More specifically, how far a student would drive to an educational institution every day (and how much parking fees if they're commuting by car). If that turns out to be a significant number, they could save money by taking online courses.

Most students find that taking Google classes saves them a lot of time since they work from home-no time is spent on regular commuting. They can even take on a part-time job if they have any spare time, so they can earn while studying. This is perfect for those looking to maintain some kind of stable income while at the same time acquiring additional qualifications through Google Classroom.

5.2 Limitations of Google Classroom

1. Complex account management

Google Classroom doesn't enable multi-domain access. Also, you can't sign in to access your personal Gmail; you need to sign in to Google Apps for Education. As a consequence, if you already have your own Google ID, managing Google Accounts can be challenging. For example, if your Gmail contains a Google document or a picture and you want to share it in the Google Classroom, you need to save it separately on the hard drive of your device, log out, and then log in with your Google Classroom account again. Pretty much trouble.

2. Too much 'Googlish.'

Google users can get confused for the first time, as there are several buttons with icons that are only familiar to Google users.

Also, despite improved collaboration between Google and YouTube, which dramatically helps with video sharing, support for other standard tools is not built-in. You can find it annoying that you need to convert a primary Word document to a Google Doc to work with, for example. All in all, in the Google Classroom environment, you'll only find yourself relaxed as long as the resources you're using fit with Google services.

3. Problems editing

When you create an assignment and send it to the learners, the learners become the document's "owners" and are allowed to edit it. This means they can erase any part of the assignment they choose, which may create problems, even if unintentionally, it happens. Also, after you edit a post, students don't get a notification.

Also, there is no direct video recording option. It would be helpful to be able to quickly and directly record voice and video messages into the Classroom at Google. Users can record the videos outside the Classroom, and then upload them as an attachment.

4. Sharing work with learners

Learners are unable to share their work with their peers unless they become "owners" of a document, and even then, they will have to accept sharing options, which will create a rift if they want to share a document with their 50 + classmates say.

5. Limited integration

Google has restricted integration options and still needs to expand these options.

6. Updates are not automated

Event feed does not automatically update, so learners have to check and regularly refresh to avoid missing important announcements.

There are both advantages and limitations of Google Classroom. Nevertheless, benefits certainly outweigh the drawbacks. Despite safeguards in place to prevent the spread of novel coronavirus, and the school year potentially canceled for thousands of students, Google Classroom is an integrating spot, considering the current state of affairs. Its usage is safe. It takes about half an hour to learn how to use it. Educators can post in an ad-hoc fashion all the essential materials, assignments, and quizzes. The software may assist private tutors, as well as home-school parents.

Chapter 6: How to get the most out of google classroom learning?

Google Classroom streamlines student-work management — announcement, assignment, selection, grading, reviews, and return. It has saved a lot of work hours for teachers. The digital job of grading can be tedious without a reliable workflow and a particular strategy. Google Classroom makes it more useful to collaborate with students — but only if you are aware of how Classroom works and how to use it to your advantage. With a few tips and strategies, Google Classroom can be even more successful in making it productive and effective.

Hack # 1: Better organized work

Use the organizational structure that is feasible for you. If your classwork page was a folder-stuffed filing cabinet, what would you place on those tabs? There are a host of systems that you can use. And thankfully, you'll be able to change your mind. Themes are too easy to modify. Themes to rename, add and remove are so simple. You can turn to a whole new organizational system in a matter of minutes if one does not fit for you.

Here are some examples you could use of the structures:

a. Classify by week

b. Classify by Unit

c. Filter by subject

d. Order by file form

Of course, you can change all of these. But having some suggestions can help you figure out which one suits you best — or at least which one you would like to try first.

Organize each subtopic more thoroughly. For example, the organizational structure can smoothly go deep into two levels—chapters and lessons. Decide how you are going to type it into the topics, and it suits perfectly. Then, stick to it. You just get a chapter and a lesson here merely by typing them in the name. Consistency keeps things perfect! It will make it much easier for students to search for the topics. They can be quicker to find what they need. With multiple levels of the organization, you don't need an official function in Google Classroom. Just create one for yourself!

Using emoji and parenthetical abbreviations as tags might be helpful. Assume you have various characteristics of your items on the Google Classroom page, for instance:

- Some might have images.

- Many of those things could be published

- Some could derive from different academic subjects

- Some might have different types (poetry versus short story versus novel)

- Others might just be interesting or funny stuff you want to share with students.

Assign an emoji in the name for any of those characteristics. Indeed, you already have an emoji keyboard on mobile devices! (If not, look at how you can connect one to your keyboard.) Use an extension like Emoji Keyboard on

Chromebooks and computers running the Google Chrome web browser.

To build emoji, use the Ctrl + Cmd + Spacebar keyboard shortcut on Macs.

If your assignment involves a photo, is a written assignment and is a social studies activity, you could use one emoji for each. To use this, think about what attributes students may be looking for in your classwork. Using an emoji lets them, in a glance, locate their task. Don't like using emoji? (Or, would you like to add a second layer of tagging?) Alternatively, try the text tags. For instance, any research project-related activity could be tagged with the abbreviation (RES).

Now, you are fully ready to take charge of a well-organized Google class.

Find a framework for the organization.

Fill in subtopics.

To add tags to classwork, using emoji or text abbreviations.

Either use one such strategy, choose two, or all three! Then watch your Google Classroom fall in line with your hopes and dreams of organizing!

Hack # 2: Reminder about the older material

To put relevant older content back to the attention of the students, use moving to the top. This simple act bumps the top of the class stream with an assignment, announcement, or question. Use this if students have not turned up a task or if you want to remind them of a deadline to come.

Hack # 3: Using the "Student" tab to email everyone simultaneously

In the "Students" tab, email each student in a class. When you click the "Students" tab, click the checkbox to highlight each student. Click "Actions" and "Email." It is nice to call special attention to what you want to communicate to students or to communicate in a longer form.

Hack # 4: Using the relevant comment type

There are various kinds of comments you can leave in the Classroom for students. Understanding how each one works will increase productivity and effectiveness.

- **Making class comments:** Insert a statement "outside" an assignment or notification into your class stream. This will make the statement available to the entire class (vital if it's an answer to a problem that everybody may have).

- **Making private comments:** Do so by viewing student results and clicking on a student individually. The comment bar at the bottom on the right, where you can see student submissions, adds a message that only the student can see (important if it has sensitive grade information or feedback).

- **Comments to be added in a doc/slide/sheet/drawing:** Do this by clicking on the student file that he/she submitted. After highlighting what you wish to comment on, press the black speech bubble button. It makes a very pointed emphasis on particular things in student work (important for input to be very exact).

Hack # 5: Sharing "right now" links using announcements

Announcements place content in the classroom stream without the need for students to hand in an assignment. Use them to provide important links, docs/files, and videos for students, which they will need right away. (If this is a resource which they would frequently use, add the resource to the "About" tab instead.)

Hack # 6: Using the keyboard, rather than the mouse

Using the keyboard commands, do away with the clicks of the mouse every time. Google Classroom's best one: When entering classes, type the grade for the assignment of a specific student, then press the down key to hit the next student. Loop with keystrokes in place of mouse clicks for students to save lots of time.

Hack # 7: Posts' reuse

Do not recreate assignments, announcements, or questions similar to those that you have already created. Press the "+" option in the bottom right-hand corner and select "reuse message." Choose your previously generated task, announcement, or request. You should update and change it before reposting it. You can also opt to make new copies of all the attachments you used before when you revisit a message.

Hack # 8: All grades in one location

In the top left of the Classroom, click the three lines button, and select "Work" at the top. Here you can find all the tasks in one place for all of your classes. Scroll down the list and get at one spot on top of everything.

Hack # 9: Get Classroom emails the way you want them

Do you spend too much time removing Classroom email messages, and wishing you could turn them off? At the top left of the Classroom, press the three lines button, and select "Settings" at the bottom. The checkbox allows you to turn off email notifications. (Or if you turn it off and wish you could get emails, that's where you turn it on!)

Hack # 10: Get others' ideas and opinions

Educators who use Google Classroom also hang out in other online forums where you could read their posts and ask questions. Here are a few suggestions:

- Google Classroom Group on Google Plus

- Mobile Applications for Education Group on Google Plus (with a Google Classroom category)

- Twitter hashtag: # Google Classroom (for Google Classroom-specific posts)

- Twitter hashtag: # Google EDU (for general Google Classroom updates)

- General Pinterest tools on Google Classroom

- A Pinterest board by Shake Up Learning

Hack # 11: Get the features Classroom needs through feedbacks

Have a suggestion for a new feature in Google Classroom? You can do something, but wish you could do something more manageable? That type of input is what Google Classroom wants from the teachers. Click on the "?" at the bottom left of the screen and pick 'Send Feedback.' According to a member of the Google Classroom team, somebody in their group reads every single feedback item sent their way. That is how they have made all of the significant improvements to the Google Classroom app. And the more frequent a request for a feature, the more likely it will be put into effect. So give reviews, and always give them!

Hack # 12: By posting a question, let the students support each other

Teachers needn't answer every question! They should provide the students with the power to help each other. To create a question for a specific task or mission, use the "+" icon. This can act as a platform of discussion, where students can support one another. (Of course, you should also look at the topic to make sure it's correct and timely.)

Hack # 13: Add learning targets

That is not as much a convenience case as it is great pedagogy. It will always remind you about the unit plan or the regular curricular, keeping you grounded.

The inclusion of the learning targets in each assignment or activity would keep reminding the students what the aim of each learning activity is.

Conclusion

Most analysts believe that online learning is taking over as the future of education. Online education shows an upward trend, especially in the prevailing year 2020. A distance learning program may be either a full distance learning or a combination of distance learning and conventional (called hybrid or blended) classroom instruction. Google Classroom is productively integrated with other Google resources like Calendar, Google Docs, Photos, Drive, and more. Educators would be able to build classes, set assignments, send feedback to the individuals, and see all features in one place. Video assignments encourage collaboration and make interactions more comfortable. Google Classroom lets students work together in several ways. Teachers may facilitate online conversations between students and teachers and assign group projects inside the Classroom. Students may collaborate, completing tasks as assigned to them. Students can also collaborate on Google Docs and can quickly share their work with teachers. Through this platform, by centralizing e-learning materials in one cloud-based location, users can go paperless and stop worrying about printing, storing, or even losing the learners' tasks. Unless you already own a personal Google Password, it can be challenging to handle Google Accounts.

Google Classroom does not allow access to multi-domain. Event feed doesn't change automatically, so learners need to review and refresh to avoid missing essential announcements periodically. Thus, Google Classroom comes with both benefits and drawbacks.

Nevertheless, the benefits far outweigh the cons. Instructors may fill in subtopics and apply tags to classwork, using emoji or text abbreviations, to find a suitable structure for a well-organized Google class. Reviews and feedbacks are critical for bringing in new functionality to the Classroom at Google platform. Teachers can use announcements to provide the students with essential links, docs/files, and videos they'll need immediately. Teachers don't need to answer every question! They should give the students the power of mutual support. Use the icon "+" to construct a query for a given task or project. It will serve as a discussion forum where students will encourage each other. Having the learning goals attached to each task or activity in Google's Classroom will keep reminding the students what each learning activity aims to be. With a few tips and strategies, Google Classroom could be made productive, efficient, and even more useful.

References

Distance education. (2020). Retrieved 2020, from **https://en.wikipedia.org/wiki/Distance_education**

Kang, T. (2020). South Korea's Coronavirus-Era Online Learning Hits Snag. Retrieved 2020, from **https://thediplomat.com/2020/04/south-koreas-coronavirus-era-online-learning-hits-snag/**

Types of Online Courses | CAS Online Education. (2020). Retrieved 2020, from **https://oe.uoregon.edu/types-of-online-courses-2/**

Classroom API overview - Classroom Help. (2020). Retrieved 2020, from **https://support.google.com/edu/classroom/answer/6253304?hl=en&ref_topic=7175285**

Google Classroom Reviews: Pricing & Software Features 2020 - Financesonline.com. (2020). Retrieved 2020, from **https://reviews.financesonline.com/p/google-classroom/**

The beginner's guide to Google Classroom. (2020). Retrieved 2020, from **https://www.bookwidgets.com/blog/2017/05/the-beginners-guide-to-google-classroom**

A Timeline of Google Classroom's March to Replace Learning Management Systems - EdSurge News. (2020). Retrieved 2020, from **https://www.edsurge.com/news/2016-09-27-a-timeline-of-google-classroom-s-march-to-replace-learning-management-systems**

8 Myths About Online Learning: The Truth Behind the Screen. (2020). Retrieved 2020, from **https://www.rasmussen.edu/student-experience/college-life/myths-about-online-learning/**

Era, 1. (2020). 15 minutes of Fame: Online Learning in the Coronavirus Era. Retrieved 2020, from **https://www.al-fanarmedia.org/2020/04/15-minutes-of-fame-online-learning-in-the-coronavirus-era/**